# Object-Oriented
# Design Choices

# Object-Oriented Design Choices

**Adair Dingle**

**CRC Press**
Taylor & Francis Group
Boca Raton London New York

CRC Press is an imprint of the
Taylor & Francis Group, an **informa** business

A CHAPMAN & HALL BOOK

First Edition published 2021
by CRC Press
6000 Broken Sound Parkway NW, Suite 300, Boca Raton, FL 33487-2742

and by CRC Press
2 Park Square, Milton Park, Abingdon, Oxon, OX14 4RN

### Library of Congress Cataloging-in-Publication Data

Names: Dingle, Adair, author.
Title: Object-oriented design choices / Adair Dingle.
Description: First edition. | Boca Raton : CRC Press, 2021. | Includes
bibliographical references and index.
Identifiers: LCCN 2020043691 | ISBN 9780367820183 (paperback) | ISBN
9780367820817 (hardback) | ISBN 9781003013488 (ebook)
Subjects: LCSH: Object-oriented programming (Computer science) | Computer
software--Development.
Classification: LCC QA76.64 .D56 2021 | DDC 005.1/17--dc23
LC record available at https://lccn.loc.gov/2020043691

ISBN: 978-0-367-82081-7 (hbk)
ISBN: 978-0-367-82018-3 (pbk)
ISBN: 978-1-003-01348-8 (ebk)

Typeset in Minion
by KnowledgeWorks Global Ltd.

# Contents

## Section II: **Strategic Type Coupling**

## Chapter 4 ■ Composition                                         89

# Preface

## WHY THIS BOOK?

A classic software engineering adage is "anyone can build a doghouse". The idea is that doghouses are not usually equipped with indoor plumbing, central heat and ventilation, and are not mortgaged, multistory or subject to building codes, etc. The list goes on. In contrast, a skyscraper must meet building codes, is likely multistory and multi-use, and, traditionally has underground parking, etc. The analogy for software development is that small-scale endeavors may be undertaken with far less overhead, and are subject to far less scrutiny than large-scale or complex systems. Design becomes more important as scale, complexity, and/or performance expectations increase.

Why then read this book? The short answer is to study software design from a structured but hands-on perspective and to understand different designs to manage types, program memory, dynamic behavior, extensibility, etc. Software complexity, refactoring expectations, and the prominence of legacy systems motivate an interest in software design. We evaluate and compare designs in this text using and contrasting C# and C++ implementations.

Software tools, standard libraries, testing methodologies, and modern IDEs have decreased the complexity of producing software. Hardware and environmental dependencies have been abstracted away. Data storage and retrieval have been streamlined. Utilities provide functions for sorting, selection, and comparison. Standard algorithms have been encoded. Yet, *one design does not fit all*. Many problems may be solved in more than one way. How does one choose the most appropriate design?

This text emphasizes design choices. Many CS texts are 'learn to' books that focus on a specific programming language or tool. When perspective is so limited, high-level concepts are often slighted. Students may gain exposure to an idea via a 'cookbook' implementation and fail to recognize foundational paradigms. Students and/or practitioners can apply principles more readily when design is explicitly defined, illustrated, and evaluated. This book analyzes competing design solutions, contrasting cost, and benefits as well and internal versus external perspectives. Expectations of code reuse trigger consideration of long-term versus short-term use. Design, not just syntax, must be stressed.

## WHO SHOULD READ THIS BOOK?

This text originated from material developed and updated for an advanced undergraduate course on software design. Students then are a natural audience. Entry-level or immediate developers, *especially those responsible for maintaining or refactoring a legacy system written in an Object-Oriented Programming Language (OOPL)*, may benefit from this explicit exposition of object-oriented design as well as the evaluation of design variants.

Some software development experience is assumed, as is knowledge of basic data structures. Expertise with any particular language, platform, or IDE is not required. Auxiliary definitions and references are noted in the text. Appendix A reinforces indirection and details relevant to C++ and C#. Appendices B and C present and analyze substantive design examples. An extensive glossary is included, defining over 150 common terms associated with software design.

## WHAT SHOULD READERS EXPECT
## TO GAIN FROM THIS TEXT?

This book provides a practical summary of object-oriented design (OOD), an emphasis on contractual design, a succinct overview of memory management and data integrity responsibilities, and evaluation of comparative design options, all promoting utility and reuse. Additionally, developers with backgrounds in either C++ or C#/Java, but not necessarily both, may benefit from explanations of language differences.

Sample design quandaries addressed are when to use inheritance (polymorphism) and when not; how to design extensible code; when to externalize dependencies; and how to simulate multiple inheritance. Each chapter ends with one or more problems which illustrate the key concepts just covered. Appendices provide solutions alongside analyses to reinforce a clear understanding of design and implementation.

## HOW IS THIS TEXT RELEVANT TO (PROFESSIONAL) SOFTWARE DEVELOPMENT?

In the rush to fill technical positions, acquisition of skills may be prioritized over concepts: learn a new programming language, use a new tool, assess a user interface in order to add functionality to an existing system, etc. A high-level perspective may be lost. Without such a perspective, software development may yield applications that are feature-rich but not easily usable or reusable. To place design into context, we uncover background processes and discuss reuse potential so that a software developer can gauge the impact of design.

## WHO MIGHT NOT BENEFIT FROM READING THIS BOOK?

Developers with extensive software design and implementation experience may find this text too elementary, unless more exposure to contractual software design or a transition to OOD is desired. Developers interested in examining real-time systems, event-handling software, or distributed systems should consult a different text.

Novice programmers may be overwhelmed. This book is not a 'learn to program' book. It can be viewed as a 'software design' book. Though many code examples are given and supportive appendices provide specific C++ and C# examples, the book has a conceptual rather than a syntactic emphasis: design (not syntax) remains the focal point. A novice programmer who wishes to learn C++ or C# should consult another text, returning to this text after mastering fundamental data structures.

# Acknowledgements

Many students, colleagues, scholars, and professional associates contributed to the form and content of this book. Student feedback led directly to foundational design examples, reworked and extended for clarity and reuse as well as contrasting implementations. I especially appreciate student enthusiasm for contractual design and curiosity about sustainable designs. Software design is not new. Experts have endeavored for years to promote deliberate and effective design. Their writings and insights have enriched software development for many years – to note a few: Michael Feathers, Barbara Liskov, Scott Meyers, Bjarne Stroustrup.

The Seattle University CS department has supported educational endeavors alongside an emphasis on professional development. Software design has thus been highlighted. I wish to thank many colleagues both in and outside US– to note a few: Renny Philipose, Susan Reeder, Roshanak Roshandel, Madalene Spezialetti, Ben Tribelhorn, and Yingwu Zhu. Special thanks to Lisa Milkowski, who read and edited several chapters. And, as always, sincere appreciation to Michael Forrest Smith for extraordinary wisdom blended with humor and optimism.

This book has been prepared with the timely and professional assistance of the editorial staff of CRC Press/Taylor and Francis Group. Thanks to Talitha Duncan-Todd for patiently guiding me and for efficiently managing all details. Repeated thanks to editor Randi Cohen for supporting and overseeing this project as well as tracking and reinforcing educational currency.

Most importantly, thanks to Tom Hildebrandt, a brilliant developer who can design or redesign anything. His early work on move semantics and his disassembler design are just two examples of his contributions to the professional software community.

# Detailed Book Outline

This book compares designs variants and emphasizes the strategic use of types in object-oriented design (OOD). In addition to thorough content coverage, many design problems are presented with sample solutions discussed in appendices. The book is partitioned into three sections that cover type *design, coupling,* and *reuse.* Eight chapters, three appendices, a glossary of over 150 common software design terms, and a list of references comprise the text. To accommodate a variety of readers, we provide sample reading suggestions.

Section I: Stable Type Design

This section reviews OOD with an emphasis on type, use of memory, and preservation of data integrity.

**Chapter 1: Contractual Design and the Class Construct** examines contractual design: software written to fulfill a contract with the client. Explicit contractual documentation with design assumptions, conditions, and invariants is reviewed. Foundational class design is defined in this context with a clear partition separating the responsibilities of the class designer and the client.

**Chapter 2: Ownership – Abstracted but Tracked** details memory management within the class construct. Allocation and deallocation are compared in C++ and C#. Examples and discussion enumerate class responsibilities when heap memory is allocated internally in a C++ object. The cost of memory ownership is considered for both languages. Storage versus computation as a primary design choice is explored.

**Chapter 3: Data Integrity** considers unwarranted aliasing as a chief cause of data corruption. Designs to avoid such problems are presented. Copying variants are analyzed, emphasizing the design responsibility to explicitly determine copy semantics. Modern constructs developed to sustain correct memory management are examined.

## Section II: Strategic Type Coupling

This section explores different ways in which to structure interdependent types, providing design examples, analyses, and guidelines.

**Chapter 4: Composition** examines the has-a relationship, analyzing deferred instantiation, echoed interfaces, and wrapped delegates as sample designs that afford significant internal control. Dependency Injection is defined and evaluated as a technique that externalizes dependencies to promote flexibility and to enable testing.

**Chapter 5: Inheritance** reviews the is-a relationship as a popular OOD choice, noting its overuse alongside analyses of costs and benefits. Language support of polymorphism, its costs, and utility via heterogeneous collections are noted.

**Chapter 6: Inheritance versus Composition** contrasts composition and inheritance as design preferences, exploring costs and reuse potential. The validity of different inheritance designs, the viability of composition, and the efficacy of using composition alongside inheritance are evaluated.

## Section III: Effective Type Reuse

This section provides a detailed evaluation of OOD variants with a particular emphasis on type reuse.

**Chapter 7: Design Longevity** considers design sustainability. Discussion includes short- and long-term assessment of abstract types, polymorphic delegates, and interface extensions. Simulated inheritance provides a foundation for many design variants. Multiple inheritances and its simulation are covered. An abbreviated production code example is analyzed.

**Chapter 8: Operator Overloading** proposes a type of design that mimics primitives via overloaded operators, permitting use in generic algorithms and containers. Again, designs are evaluated for language differences and consistency.

## Appendix A: The Pointer Construct

This appendix covers the 'pointer' type, a language construct provided in C and C++, but not in C# or Java. Proper use as well as inappropriate handling is illustrated through pertinent examples. With its thorough coverage of indirection, Appendix A assists the C# or Java programmer who is learning C++.

## Appendix B: Design Exercises

This appendix provides sample design solutions, in both C# and C++, to problems posited at the end of Chapters 1 through 5. Illustrative problems and contractual designs are presented for class design, proper class memory management, copying, composition, and inheritance.

### Appendix C: Comparative Design Examples

This appendix provides sample design solutions, in both C# and C++, to problems posited at the end of Chapters 6 through 8. Contractual designs are presented that compare inheritance versus composition, type reuse for longevity, multiple inheritance, and operator overloading.

### Sample Readings:

To accommodate different levels of experience, sample suggested readings of the text are given below. Regardless of experience, Appendix A covers material highly recommended for readers without a C or C++ background. Appendices B and C serve to illustrate stable OO designs and to support design discussions.

Intermediate Programmers or Python/Ruby/Javascript Programmers:

the entire text is applicable -- Chapters 1-8, Appendices A, B, C

C Programmers Transitioning to OOD:

Chapters 1, 3-8, Appendices B, C

C# or Java Programmers Familiar with OOD but Transitioning to C++:

| | |
|---|---|
| *Language Impact:* | Appendix A, Chapter 2, 3 and 8 |
| *Comparative Design:* | Chapter 6, and 7, Appendix C |
| *(OOD Review:* | Chapter 1, 4, 5) |

### Chapter Format

The acronym SOLID traditionally summarized five principles of OOD: Single responsibility principle, Open closed principle, Liskov substitutability principle; Interface segregation principle, and Dependency inversion principle. This text covers these essential OOD principles and more. Each chapter is associated with at least one principle. The chapter's design exercises are meant to highlight one or more relevant principle and are evaluated in that context in the corresponding Appendix.

Each chapter begins with a bulleted list of chapter objectives. Extensive code samples, design examples, figures, and summative tables augment the prose. Common software terms (defined in the glossary) are bolded upon first use as are emphasized ideas. Italicized comments highlight design principles or key insights.

Each chapter ends with the same structure: definition of a relevant design principle, the chapter summary, one or more sample design problems (with solutions sketched in the noted Appendix), design insights, and conceptual questions intended to review major concepts.

## Code Samples

All code samples were compiled and run. Visual Studio19 processed C# code. A gnu C++17 compiler processed C++ code. For brevity, common declarations (such as libraries and namespaces) were omitted when the code was copied into the text. For example, in C++, "`using namespace std`" was commonly omitted as was "`#include    <iostream>`". However, less common inclusions, such as "`include <algorithms>`", were noted when used. C++ client code was assumed to execute in main. Similarly, in C#, common libraries were not noted, nor was the wrapping of client code in a class construct.

# I

## Stable Type Design

## CHAPTER 1

# Contractual Design and the Class Construct

## CHAPTER OBJECTIVES

- Outline contractual design

- Define Programming by Contract

- Examine standard class components

- Identify relevant OOD principles

## 1.1 ENCAPSULATION

Edsger Dijkstra, a computer science pioneer, famously observed "simplicity is prerequisite for reliability". A simple design is easier to maintain than a complex one. Traceable control flow is easier to understand, and modify, than tangled branching. An elementary type is easier to manipulate than a complex one, etc. Much of design thus strives to make components simple, or appear to be so. **Object-Oriented Design** (OOD), when used appropriately, promotes simplicity by partitioning type definitions: internal (implementation) and external (interface) perspectives distinguish functionality from use.

OOD dominated software development in the 1990s and early 21$^{st}$ century. A design approach that rests on the notions of **abstraction**,

**encapsulation,** and **information hiding,** OOD supported the development of large-scale software systems and advanced the concept of **code reuse. Design Patterns** standardized common solutions to reoccurring problems and popularized OOD design principles.

OOD encapsulates private data, providing public methods for the client without direct access to encapsulated data. Class (type) definitions isolate the client from internal details, ensuring type consistency. OOD thus sustains the high level of abstraction needed to build and maintain large software systems. Many **legacy** systems are written in object-oriented programming languages such as C++, Java, and C#. To maintain and refactor legacy code for continued use, one must understand the structure and effect of OOD.

Intentional design builds on key object characteristics, such as **lifetime, association, ownership,** and **cardinality,** yielding insights that are transferable to higher level views of software. Are objects temporary or persistent? Can object instantiation be postponed? Is the association between two objects permanent or transient? Who owns a subordinate object? Can ownership be transferred or shared? How many objects exist in a relationship? Is that number fixed? Other questions arise. *Deliberate design must identify assumptions.*

*Compilers do not enforce design!* Maintainable, extensible code is developed by documenting assumptions, following established design principles, and choosing appropriate design variants. But the compiler will not verify design merit; its task is to follow a long, complex set of instructions for source code translation.

## 1.2 EXPLICIT DESIGN AND CONSTRAINTS

OOD starts with type definitions, where every type supports external and internal perspectives. Each object instantiated from a type definition has its own internal values and **state.** A client manipulates types from an external view, invoking public functions as needed. A class designer implements types with an internal perspective, defining properties, and preserving state control. OOD becomes more complex when reusing types in different structural and behavioral relationships.

Consider containers that provide functionality for data storage. The classic stack data structure defines functions to add and remove data as well as state queries. A client must know whether a container is in an empty or non-empty state. Functions such as isEmpty() operate in the same manner regardless of the data type stored in the stack. Structurally then, as is true for most containers, a stack looks the same regardless of what type of data it holds. Functional independence from data type is implicit for standard containers such as queues, priority queues, trees, etc. A generic design thus promotes consistent use (and reuse) of containers.

In contrast, type impacts data classification systems – commercial inventory, library holdings, university courses, etc. A basic definition may isolate common features of classified items, such as quantity, date, etc. Specialization adds detail and functional variation. For example, restocking games in a toy store differs from returning books to a library. To promote consistent use (and expansion), design rests on a type framework that permits variation.

Containers and classification systems illustrate different dependencies. Containers are type agnostic: primary tasks of storing and retrieving data are implemented without much regard to type. Classification systems are grounded in type: functions that verify, replace, order, etc. items rest on type.

## 1.2.1 Class (Type) Functionality

The class construct formalized the implementation of an **abstract data type (ADT)**, and, in so doing, streamlined the idea of encapsulating data alongside functionality. The systematic design of a class generalizes to software design. Functionality can be delineated by intent: initialize data; allocate, deallocate, or manage resources; change or view data values; and, examine or resolve data dependencies. Fulfillment of functionality may be conditional: a request for access or change may be denied. Table 1.1 categorizes standard functionality defined within a class: constructors,

TABLE 1.1 Types of Functions Defined in Class Construct

| | Intent | Invocation |
|---|---|---|
| **Constructor** | Initialize data | Object instantiation |
| | Allocate or acquire resources | *Note language differences* |
| **Accessor** | View data values | Depends on accessibility |
| **Mutator** | Change data values | Depends on accessibility |
| | Preserve validity of state | |
| **Public interface** | Support type definition | Unrestricted access |
| | Provide needed utility | client, type or subtype |
| **Private utility** | Preserve data dependencies | Internal |
| | Manage resources | |
| **Destructor (C++)** | Release resources | Exit scope (automatic) |
| | Bookkeeping | Explicit call to delete |

destructor, accessors, mutators, private utility functions, and public interface functions. Functions defined within the scope of a class are often called **methods**.

Each method in a well-designed class supports abstraction and encapsulation and, as much as possible, allows the client to treat the custom type as if it were a built-in type. Initialization and preservation of state are internalized and not left to the client. Hence, it is common to differentiate between private methods that may be called only from within other class methods and public methods that the client can directly invoke.

Example 1.1 provides C++ code for a sample Icon class which represents a visual element in a computer game. Example 1.2 shows comparable C# code. This incomplete design is not immediately usable but illustrates a type definition. Sample data fields display object form, and sample methods represent movement and/or change within a game session. A computer game would be seeded with multiple Icon objects, each potentially in different states.

### Example 1.1  Sample C++ Class Design

```cpp
class Icon   // C++ data members and functions,
             //              private by default
{      // data allocated for each
       // instantiated Icon object
       double           mass, energy;
       int              x, y;
       // dependent on energy
       bool             visible;

       // static data: ONE per class;
       // shared by all objects
       //C++17 in-field initialization

       inline static int          count = 0;

       // private utility method
       void adjustEnergy();
   public:
       // constructor must set state, that is,
       //initialize fields and increment
```

```
//    static count: one more Icon object
Icon(int xC = 0, int yC =0)
{    x = xC;              y = yC;
     mass = abs( x * y) + 10;
     energy = abs( x + y) + 100;
              // invariant
     visible = mass < energy;
     count++;
}

// destructor decrement count:
// one fewer Icon object
// no resources to release
~Icon() {  count--;  }

// accessor: control view; may choose
// NOT to return value
double     getEnergy() const
{    if (visible)              return energy;
        return                 -1;
}

bool   isVisible() const
{ return  visible;  }

static int  getCount()
{ return  count;     }

// mutator: controls state;
// may reject change request
bool refuel(double     fuel)
{    if (!visible || fuel < mass)
          return false;
     energy += fuel;
     return true;
}

// may change visible
void  flicker()  {    ...    }
   ...
};
```

The Icon class defines private data: integers x and y to model grid placement, doubles mass and energy to control movement and Boolean visible to control fulfillment of client requests. Each instantiated object has its own copy of these data fields. A static integer (count) tracks the number of objects instantiated from the class. Static data members are allocated at the class level; all objects share a single copy. Static data members are initialized in the class definition for C#, Java, and C++17 but initialized in the .cpp file for legacy C++.

### Example 1.2  Icon Class in C#

```
public class Icon
{       private     double      mass;
        private     double      energy;
        private     int         x;
        private     int         y;
        private     bool        visible;

                    // static data: ONE data
                    // member PER class
        private     static int count = 0;

        // private utility method
        private void adjustEnergy();

        public Icon(int xC = 0, int yC =0)
        {   x = xC;         y = yC;
            mass = Math.abs( x * y) + 10;
            energy = Math.abs( x + y) + 100;
            // invariant
            visible = mass < energy;
            count++;
        }

        //not really a destructor:
        //called before object reclaimed
        // no resources to release
        ~Icon() {   count--;    }

        //C# property: get (accessor) and set (mutator)
```

```
public bool Visible          //replaces isVisible
{    get => visible;   }

// Property replaces getEnergy
public double Energy
{    get   {      if (visible)    return energy;
                  return              -1;
      }
}


public static int    getCount()
{ return       count;        }

// mutator: controls state;
// may reject change request
public bool refuel(double   fuel)
{    if (!visible || fuel < mass)
         return false;
     energy += fuel;
     return true;
}

public void      flicker()
{   // may change visible' … }
   …
}
```

## 1.2.2 Constructors

Constructors are special methods that share the class name and return no value after initializing an object's data members. The compiler patches in a call to a constructor when an object is instantiated, because initialization is a class responsibility, not the client's. The provision of a public `initialize()` undermines class control: at any time, an object could be reset to an initial state. Since constructors are called only once, there is no vulnerability to such an arbitrary reset. Constructors may be overloaded, that is, more than one may be defined, each distinguished by its parameter list.

An object definition is a two-step process: memory is allocated and then a constructor fires to initialize data fields and put the object into an initial, valid state. When objects are declared in C++, the compiler automatically patches in a call to the **no-argument constructor**. The terms **default constructor** and no-argument constructor are often interchanged and thereby confusing. Default constructor is an older, C++ term and refers to the constructor provided by the compiler if the class designer does not define any constructors. The default constructor never takes any arguments. How could the compiler decide what arguments to pass? Java and newer languages refer to the no-argument constructor as the constructor (defined or default) that takes no arguments.

In Java and C#, object declaration allocates only a typed reference (which holds the address of an object after its allocation via a call to the new operator). Calls to the new operator explicitly identify the constructor to invoke. Similar to C#, a typed C++ pointer may be used to hold an address of memory allocated by a call to the new operator. Object allocation and management is discussed in more detail in Chapters 2 and 3.

**Example 1.3  C++ object Instantiation**

```
// #1 no-argument constructor
Icon              objX;
// #2 overloaded constructor
Icon              objY(10);
//#3 no-argument constructor
Icon*             ptrX = new Icon;
//#4 overloaded constructor
Icon*             ptrY = new Icon(3);

// C++ array declaration: no-arg
// constructor always invoked
// all objects in array initialized
// to same initial state
// #5 100 calls to no-arg constructor
Icon        db[100];

// overwrite default initialization
// of array elements
for (int j = 0; j < 100; j++)
{
        // #6 constructor takes int
        Icon    replace(j);
```

```
            // #7 overwrite array entry
            db[j] = replace;
    }
```

Example 1.3 illustrates different object instantiations in C++: statements #1, #3, and #5 invoke the no-argument constructor; statements #2, #4, and #6 invoke the overloaded constructor that accepts an integer parameter. In C++, when an array of objects is allocated, the compiler patches in multiple calls to the no-argument constructor, one call for each array element. What if the objects in an array should be 'constructed' (initialized) by a different constructor? Extra code must be inserted to overwrite array elements, as shown in statements #6 and #7.

All C# objects are allocated via a call to the new operator so the client must specify a constructor. An array of C# objects is really an array of references, where each reference should be assigned the address of an object. Initializing an array of C# objects thus always requires stepping through the array of references to allocate the objects referenced therein. Example 1.4 illustrates different object instantiations in C#. Statements #1 and #3 invoke the no-argument constructor; statements #2 and #4 invoke the overloaded constructor that accepts an integer parameter. Statement #5 allocates an array of references on the heap; no objects are allocated. Individual objects are allocated in statement #6, a process that easily supports custom invocation of a constructor.

**Example 1.4 C# object Instantiation**

```
        // C# object declaration: Icon reference
        //    no object allocated yet;
        //    reference zeroed out
        Icon objA;

        // C# object declaration and instantiation
        Icon  objB = new Icon();              // #1
        Icon  objC = new Icon(12);            // #2
        ...

        // C# object instantiation
        // overwrites references
        objA = new Icon();                    // #3
        objC = new Icon(15);                  // #4
```

```
// C# array declaration:
// array of 100 Icon references
Icon   db[] = new Icon[100];                  // #5

// C# array initialization:
//      each reference (array element)
//             holds address of object
// allocated by new
for (int j = 0; j < db.Length; j++)
     db[j] = new Icon(j);                     // #6
}
```

*Constructors are responsible for setting the initial state of an object,* which may include allocating or acquiring resources. As illustrated in the Icon class, data members may be initialized by a default value, a parameter, or another data member.

### 1.2.3 Accessors and Mutators

Accessors, often named with a prefix of get, or in C#, defined via a property, provide a controlled peek inside an object, returning the value of a private data member. In containers, a common query is get-Size(), which returns the current size of the data set. No accessor should modify the state of an object. Hence, C++ accessors should be labelled const. Accessors may check state before returning a value, and may reject a request for data, as shown in Icon::getEnergy(): if the object is invisible, -1 is returned as an error code rather than the actual energy value.

Mutators, often named with a prefix of set, or in C#, defined via a property, allow the client to potentially alter state by changing an internal value. To preserve internal control, set() functions may be conditional. A set request may be rejected if the value provided would put the object in an invalid state. Common examples of refused requests include out-of-bounds values or values that violate a dependency between two or more fields in the class definition. The mutator method Icon::incEnergy(x) rejects the change request if the Icon object is invisible or if the passed value x is less than the mass data field. A class designer preserves integrity with externally inaccessible (private) data alongside conditional mutators that preserve type consistency.

C# properties are easily defined in modern IDEs but should not be overused. Class designers should ensure that only appropriate accessors and mutators are provided to the client, and make fulfillment conditional if necessary. *Accessors do not change state*, as illustrated by const C++ methods in Examples 1.1. *Mutators alter state in a controlled manner*, as illustrated by incEnergy() in Example 1.1. Class methods should not give external access to private data, which would permit uncontrolled change of state, unless the provision of such access is a deliberate design decision.

**Example 1.5  C++ Aliasing Undermines Encapsulation**

```
// mutator that rejects out of range values
void myType::setValue(int    x)
{    if    (x > 100)   hiddenInt = x;      }

// standard accessor: data member
// returned by value
int  myType::getValue() const
  {  return  hiddenInt; }

// name 'get' implies accessor
// but a reference returned
// #1
int& myType::getControl()
{ return hiddenInt; }
...

// Client code
myType       insecure;

// change request OK
insecure.setValue(200);
// change request rejected
insecure.setValue(13);
cout << insecure.getValue() << endl;    // #2

int&   alias = insecure.getControl();   // #3
cout << insecure.getValue() << endl;    // #4

// #5 private data member altered
alias = 13;
cout << insecure.getValue() << endl;    // #6
```

Example 1.5 illustrates an accessor that compromises object state by aliasing an external data variable to an internal data member. getValue() is labelled const while getControl() is not. Yet, it still may be unclear how data integrity is undermined. The accessor getControl()returns an integer value by reference (statement #1). Hence, the caller's integer variable alias is aliased, or shares the same memory space, as the private data member hiddenInt (statement #3). Subsequent changes to alias change the private data field hiddentInt of the object insecure.

Contrast output values from statements #2, #4, and #6. Since the request insecure.setValue(13) was ignored, each output statement should print '200'. Yet, statement #6 outputs '13'. Why? The assignment to alias in statement #5 results in an unseen alteration of the hiddentInt data member of object insecure. This example is in C++ but aliasing (and data corruption) is possible in any language.

### 1.2.4 Utility and Public Methods

Private utility methods support functional decomposition and reuse within a class, reducing code complexity. For example, resize() expands the size of a container when capacity is reached, and may be called from any method that adds an element to the data set. Why should resize() be private? The client should not be responsible for maintaining the container in a usable condition. If capacity is unbounded, the client should not have to track available storage. When overflow is imminent, resizing should be internally triggered. The client uses services provided by a container to store and retrieve data, but should not manipulate internals. See Example 1.6.

**Example 1.6 Private Utility Method `resize()`**

```
// C++ resize() doubles capacity -
// defined in .cpp file
// declared as private in .h file
void container::resize()
{          storedType*
           temp = new storedType[2*size];
           for (int j=0; j < size; j++)
               temp[j] = heapData[j];
           size *= 2;
```

```
            delete[] heapData;     // release old data
            heapData = temp;       // reset pointer
}

// C# resize() doubles capacity -
// defined in class definition
private void resize()
{            storedType[]
             temp = new storedType[2*size];
             for (int j=0; j < size; j++)
                 temp[j] = heapData[j];
             size *= 2;
             heapData = temp;       // reset reference
}
```

OOD restricts exposure of implementation and reduces client responsibilities. Public methods should not require knowledge of internal form. For example, a client need not know or care how a container holds its data. Arrays and lists are common structures for data storage but neither choice should directly affect client code. Likewise, when manipulating Icon objects, a client should not have to manage encapsulated data: the dependency of visible on mass and energy is internally controlled.

### 1.2.5 Destructors

The C++ compiler implicitly invokes the destructor when an object goes out of scope. Like constructors, a destructor is a special method that returns no value and its name is simply that of the class preceded by the special symbol '~'. Destructors release resources that an object acquires at run-time. For example, destructors could release any file opened and used internally by an object. More commonly, if an object dynamically allocates data, a C++ destructor ensures deallocation. Chapters 2 and 3 provide more detail on memory allocation and deallocation as well as potential data corruption.

In C#, memory acquired at run-time is released 'automatically' via garbage collection; Chapter 2, examines such implicit deallocation. C# classes then need not define a destructor. However, C# permits the definition of a *finalizer* method that syntactically looks like a destructor, which is called implicitly by the garbage collector when an object is reclaimed. The utility of C# finalizers is limited since the garbage collector usually runs outside program control. In contrast, destructors are essential in C++. Standard

C++ design guidelines suggest that they always be defined. Example 1.1 illustrates a C++ destructor, ~Icon(), which will be invoked whenever a stack-allocated Icon object goes out of scope or a heap allocated Icon object is deleted. Example 1.2 shows a C# finalizer, ~Icon(), which is invoked only when the garbage collector runs and reclaims an unused ('garbage') Icon() object.

Destructors may be used for bookkeeping. Tracking the number of objects is a technique employed in resource management as well as debugging. To do so, every constructor increments a static count upon object instantiation and the destructor decrements the count upon object deallocation. The Icon class employs this design, allowing a game designer to track the number of allocated Icon objects. Since this static count is private, as it should be, a static accessor function is needed, as shown in Examples 1.1 and 1.2.

## 1.3 DESIGN AS A CONTRACT

To ensure consistent use, and to correctly manage dependencies, design must be recorded. Yet, inline comments interfere with readability and are less likely to be updated when software is modified. As noted by Ron Jeffries (one of three founders of Extreme Programming (XP) methodology), "Code never lies. Comments sometimes do". A distinction must be made between comments and design documentation. Inline comments should be avoided. Through judicious choice of names, and adherence to control flow conventions, etc., code should be readable on its own. Code should be self-documenting. At a higher level, *design documentation identifies intent and implementation details in support of software evolution.*

Tools and conventions abound for standardizing documentation. Contractual design is a documentation methodology that outlines design decisions and client responsibilities. Documentation as a contract supports both the client who uses the provided public interface and the class designer who implements the functionality to support published expectations. Introduced by Bertrand Meyers, the architect of the object-oriented language Eiffel, Design by Contract embodies a professional perspective on software development. Programming by Contract is an academic rendering of Design by Contract that uses blocked comments to publish requirements and to record design decisions. Microsoft advances a similar perspective, code contracts, where documentation serves as a contract between the client and the class designer. If the client adheres to specified restrictions, then objects should behave as expected.

## 1.3.1 Error Handling

**Defensive programming** assumes that defined types may or may not be used correctly and, thus, extensive testing is mandated to prevent error. In contrast, Programming by Contract outlines a formal agreement between class designer and client. The contract identifies requirements to be met by both parties for safe and consistent use of a defined type. By specifying shared responsibilities, a contract clarifies outcomes and may alleviate the overhead of excessive error checking. Note though that the cost of contract violation cannot be too high. When a client breaks a contract and receives invalid data but there is no other penalty, the class designer may not be concerned. However, if a broken contract may lead to data corruption (without internal checks), then the class designer must exercise caution.

The stack data structure highlights the value and limits of contractual design. A standard precondition for pop() is that the stack object is not empty. What if a client erroneously requests data from an empty stack? A defensive approach internally checks state and rejects invalid pop() requests if the stack is empty – a safe but costly approach since every caller pays for the check. Design may be tricky. How does pop() communicate a rejected request? One solution is to return an error code indicating failure when the stack is empty. Error codes cannot use any value (such as zero) that legally could be stored. Another option is to return the popped value through a parameter passed by reference, using a Boolean value as the return type to indicate validity: true if pop() succeeds; false if it fails (and thus no change to the parameter passed by reference). The most secure approach is to use exceptions.

Assume a stack holds data in an internal array and that the stack is empty when the client calls pop(). C# and Java perform bounds check-ing on each array access so an exception is thrown when the invalid index of an empty stack is used. No explicit internal check is then needed but contractual design should note that the client must be prepared to catch an exception if pop() is called when the stack is empty. In contrast, C++ does not perform bounds checking so the invalid array index would be used, triggering an exception if the referenced memory lies in a protected region, or, most likely, the return of invalid data. More insidiously, data corruption may result from this C++ scenario. Without bounds checking, if a push() follows an invalid pop(), the invalid C++ array index may overwrite memory outside the internal array. The consequences of data corruption are too severe for a C++ implementation to eschew both excep-tions and defensive programming.

When does contractual design offer a reasonable alternative to (excessive) internal checks of defensive programming? When the cost of contract violation is not too high. An error response of a thrown exception is an acceptable risk. Data corruption is not. Exceptions preserve system integrity, incurring overhead only if thrown. Therefore, exceptions do not degrade performance. Examples 1.1 and 1.2 provide a benign example of contract violation. If the client does not meet `visible` expectations for checking the energy reserves of an `Icon` object, then an error code is returned rather than accurate data. However, the integrity of the `Icon` object is preserved.

### 1.3.2 Published Assumptions

*Contractual design specifies client responsibility.* Stated preconditions must be met. Otherwise, the client breaks the contract with the class designer and behavior is not guaranteed. Preconditions must be verifiable. The client should not have to count additions and removals to determine whether a container is empty or not. A query method, such as `isEmpty()`, must be provided to allow the client to verify state.

Programming by Contract delineates obligations across five documentation categories, as noted in Table 1.2: function **preconditions**, function **postconditions**, **interface invariants**, **implementation invariants**, and **class invariants**. Pre and postconditions identify dependencies and assumptions about the environment in which functions execute. For correct execution, the client must meet preconditions before invoking a function. The class designer guarantees postconditions so that the client can track state, in order to satisfy precondition(s) of subsequent function calls. Encapsulation promotes data integrity, potentially reducing error checking. Pre and postconditions should be published, when relevant, whether or not software is object-oriented.

Preconditions enumerate conditions require for correct execution, such as a container must not be empty for data extraction or an object must be

TABLE 1.2   Programming By Contract

|  | Intent | Characteristics |
|---|---|---|
| **Precondition** | Safe entry into function | Required incoming state |
| **Postcondition** | Identify state changes | Possible altered state |
| **Interface Invariant** | Promote consistent use | Services supported |
| **Implementation Invariant** | Software maintenance | Design specifications |
| **Class Invariant** | Communicate Type & Use | Designed functionality |

'on' to enact change. The client fulfills preconditions so the function can avoid the overhead of verification. *Preconditions identify potential error(s) and minimize internal error checking.* Preconditions must be verifiable and must be published, typically as a comment preceding the function header. Callers must recognize the potential for error, or the need to catch exceptions, if preconditions are not satisfied. Preconditions may restrict acceptable values for passed parameters but need not specify type (in a statically typed language) because the compiler checks type. In a dynamically typed language, like Python, a precondition may need to specify types for which the operation holds.

Evaluate the stability of preconditions, especially for resource management since resource use may span multiple actions. Consider file existence: a check before opening a file seems secure but the file could be deleted after the existence check and before reading from the file. What designs are viable? Access could be wrapped in an exception block. Locks could be introduced. ***Contractual design identifies requirements and impediments before coding begins.***

Postconditions publish the effect of function execution, identifying actual as well as *potential* changes, such as resource acquisition or release. Postconditions do NOT describe a function's action. In the OO paradigm, postconditions identify possible change to object state after method execution, for example stack is non-empty after push(), stack is empty after clear(), stack may be empty after pop(). With postconditions, the client can track change and verify preconditions for subsequent function calls. See Table 1.3.

Preconditions specify requirements for correct use of a function, including any parameter restrictions and required object state. Programming by Contract highlights the shared responsibility between the caller and the callee. If the required preconditions are not met, there is no guarantee about resulting behavior. Postconditions record actual and potential change so that state may be tracked. Pre and postconditions support intentional design whether the function is declared

TABLE 1.3  Common Pre- and Postconditions

|  | State | Resource | Data | Ownership |
|---|---|---|---|---|
| *Precondition* | Non-empty | Memory allocated | Values in range | Valid Handle |
| *Postcondition* | Empty | Memory allocated | Ordered | Assumed |
|  | May be empty | Memory released | Unique | Released |
|  | Non-empty |  |  | Unaffected |

public, private, or protected in a class, or, in fact, whether the function is encapsulated in a class at all.

### 1.3.3 Invariants

All class member methods should ensure a valid state (legal values of data members). Constructors create objects in a valid, initial state. In C++, destructors release resources and track bookkeeping details. Accessors return copies of data values (or, with risk, aliases to internal data) without any state change. Mutators change state while preserving validity. Private utility functions provide functional decomposition within a class design. Additional methods implement core functionality.

Invariants serve to document stable state conditions. Invariants describe design decisions in the context of class structure, noting conditions and relationships to be preserved. Interface invariants are external constraints. Implementation invariants are internal constraints. Class invariants are the intersection of interface and implementation invariants – of interest to both the client and the class designer.

Interface invariants provide an overview of public use and inform the client of constraints. Interface invariants provide a higher level of abstraction than preconditions, and describe general restrictions on the use of objects. For example, no meaningful response from an 'off' sensor; no changes processed for a 'closed' inventory, etc.

Implementation invariants record all relevant design choices, in detail sufficient for software maintenance. Common decisions include: choice and expected use of subordinate data structures, legal values of data fields, ownership responsibilities, dependencies between fields, memory (resource) management, and bookkeeping details. Reasonable implementation invariant examples are: internal (static) registry guarantees unique id; data is not ordered; interface of subobject echoed – details that the client need not know. By identifying data and design constraints, implementation invariants record how requirements are met.

Class invariants represent the overlap of interface and implementation invariants that is type design decisions that constrain both client expectations and class designer responsibilities. Class invariants may specify: error processing; copy semantics; capacity limits (if any); ordering criteria (if any); processing of duplicate values (if any); etc. All operations should be designed to preserve class invariants. *The closed nature of a class gives the designer complete control over all operations that modify data fields.*

TABLE 1.4    Common Invariants for a Container

| | Utility | Data | Constraints | Error Handling |
|---|---|---|---|---|
| *Class* | Storage | Characteristics (unique, sign,,,) | Capacity Copying | Error Codes Exceptions |
| *Interface* | Type Meaning | Validity | Access | Response |
| *Implementation* | Functionality Data members Algorithms | Dependencies Stability Ownership Ordered (or not) | Resources Data structures Algorithms Relationships | NOP Recovery Replacement |

*Programming by Contract supports software maintainability by explicitly recording design assumptions*: implementation priorities and design choices must be clearly specified. See Table 1.4 for sample invariants for a container. When a class design must be modified, the software developer must reexamine the class and determine how to incorporate additional or altered functionality. This task is more easily accomplished when original design and intent are clearly documented.

## 1.4  PROGRAMMING BY CONTRACT EXAMPLE

The class designer should enumerate minimal restrictions in a contract, recording only restrictions internally enforced. For example, if one client does not process duplicate values but another client may then data uniqueness is not an internal constraint. To illustrate, we specify a contract for the cyclicSeq class sketched in Example 1.7. Encapsulation supports invariants, properties that always hold, ensuring that objects remain in a consistent, legal state (which can be tested at any point). Preservation of invariants may reduce error checking and promote software maintainability by limiting dependencies on the client. Two benefits are: 1) the class may be treated as abstractly as a built-in type since the client need not know internal details; and, 2) the design is secure because an inattentive or malicious programmer cannot easily put an object in an invalid state.

### Example 1.7  C# cyclicSeq class

```
// a cyclic arithmetic generator
public class cyclicSeq
{       // C# data zero initialized
        //   by default
        private bool        on = true;
        private uint        place;
```

```
private uint            period;
private readonly uint   a1;
private readonly uint   dst;
...
public cyclicSeq
(uint start, uint inc, uint lgth = 100)
{       a1 = start;
        dst = inc;
        period = lgth;
}

public int nextNum()
{       if (!on)      return -1;
        place = (place + 1)% period;
        return (int)(a1 + dst*place);
}

public bool expand(uint scale)
{       if (!on || scale == 0) return false;
        period *= scale;
        return true;
}

public bool isOn()
{     return on;              }

public bool toggleOn()
{     return on = !on;          }

}
```

Example 1.7 defines a generator that yields a number upon request. The number returned is the next number in a cyclic sequence, with its first value, additive factor, and sequence length defined in the constructor. The starting and increment values must be specified but a default value may be used for sequence length. A client may change a cyclic-Seq object in only two ways after construction: toggle the on state and expand the length of the sequence. The on state controls the validity of data returned as well as the ability to expand the sequence length. Sample contractual documentation for this example distinguishes *minimal*, *unnecessary*, and *problematic* details. Invariants need not document restrictions enforced by the compiler. Hence, it is <u>unnecessary</u> to state that negative values are unacceptable for a parameter typed as an unsigned integer, etc. Problematic details are those inconsistent with the type definition.

**Class invariants provide an overview of the defined type and expected use.** Minimally, type functionality must be described alongside error processing. Class invariants summarize the overlap of interface and implementation invariants – details that affect the client and that must be tracked by the class designer, such as the uniqueness of data values or the suppression of copying. Restrictive characteristics that may arise only through client actions should not be specified. **Sample class invariant** for `cyclicSeq`:

1. *Minimal details needed to define `cyclicSeq` type*

   - Arithmetic generator yields next number in cyclic sequence upon request
   - Constructor sets starting value, additive factor, and sequence length
   - Default value defined for sequence length
   - Starting value and additive factor stable throughout object lifetime
   - Sequence length may be changed upon request after construction
   - on state controls return of valid data and expansion of sequence length
     → Client must track on state
   - Error processing
     - If next number requested when not on, -1 returned as an error code

2. *Unnecessary details implied by type definition*

   - Values returned will repeat
   - Values non-negative

3. *Problematic details that have no internal validity*

   - No prime numbers – no such filter in class

**The interface invariant summarizes intended use by the client.** Initial and subsequent state requirements must be described. Comments on use or effect must be consistent with design. There is no need to describe error conditions that do not occur, for example arithmetic miscalculations or invalid initialization. What states ensure validity? What is the client's responsibility for tracking state? **Sample interface invariant** for `cyclicSeq`:

1. *Minimal details needed to specify correct use of* `cyclicSeq` *type*

    • Client must track on state

2. *Unnecessary details implied by the interface*

    • State may be toggled

3. *Problematic details that described unsupported behavior*

    • Starting value or additive factor may be modified

**Implementation invariants document the class designer perspective,** providing detail on internal design. Minimally, key design decisions must be described as well as error response. Explanations may be provided for details evident through self-documenting code, such as `uint` restricts `cyclicSeq` values to non-negative integers. Details are often language specific. **Sample implementation invariant** for `cyclicSeq`:

1. *Minimal details describe internal data structures and dependencies*

    • Values in sequence not stored, computed upon request

    • on state controls response to client requests

2. *Unnecessary details implied by function prototype (syntactically evident)*

    • only non-negative integers generated

3. *Problematic details inconsistent with internal response*

    • values may not be single digits – no such expectation established

**Preconditions notify the client of the pre-requisites of a legal call,** specifying any required state(s) before function entry. For example, containers must be non-empty before any data is extracted and must have sufficient capacity for data insertion (or provide internal resizing). The `cylicSeq` class design uses on to control behavior. Hence, a valid precondition may be 'object is on'. Again, preconditions need not state conditions enforced by the compiler, such as parameter type.

**Postconditions document potential or actual state changes so that the client may track state.** The client may need information about object state or resources in order to fulfill preconditions. Hence, the client must be able to query state since encapsulation hides data. To know when to query state, the client must recognize the potential for state change. const methods do not have postconditions because they do not change object data and thus cannot affect state. Postconditions do not describe what functions do.

Systematic documentation distinguishes form and use, separating structure from interface. More than an arbitrary convention, Programming by Contract delineates shared responsibility and records software design. It emphasizes the encapsulated design of objects, with separation of public and private interfaces, and summarizes the dual perspectives of client and class designer.

## 1.5 CONTRACTUAL EXPECTATIONS

A contract outlines responsibilities from both internal and external perspectives. In contrast, defensive programming makes no assumptions about correct input or fulfillment of preconditions and may require extensive error checking. When class methods check arguments and object state, the overhead is borne by all callers. This expense may be acceptable for infrequently invoked methods but becomes expensive for repetitive calls. Class designers should evaluate internal versus external costs.

Encapsulation supports the development of reliable code because internal state is controlled: the client cannot put an object (of a well-designed class) into an invalid state. Ideally, the class designer delineates public and private functionality so that objects are always initialized correctly and no method modifies internal data inappropriately. With controlled and checked modification, objects should always be in a valid state.

Internal responsibilities lie within a class. Control of object state should be internalized. A client may undermine design if allowed to unconditionally change object state. Why? A client may not recognize design intent, internal dependencies between data members, or restrictions on state changes. In a well-designed class, object data members are private and only the execution of member functions may change state.

The tradeoff between reliability and efficiency is explicit: error checking versus assumptions of correct use. The overhead of an internal conditional test is borne by all requests. Contractual design may be a viable alternative. The class designer should estimate consequences of contract violation, that is, the impact of error. What failures are acceptable and how are they handled? How often is error expected? The cost of failure cannot be too large. Exception handling provides a reasonable response to error. As a safeguard, *exceptions incur no run-time overhead unless an exception is thrown*. Exceptions can be used in both defensive programming and contractual design. The client should be informed of error codes and/ or the need to catch thrown exceptions. Critics complain that exception handling clutters code but many error conditions (such as resource acquisition) generate exceptions that cannot be checked via if statements.

Contractual obligations, as enumerated by Programming By Contract, help refine design decisions, including selection from different designs. This text examines and contrasts various designs, exploring costs and benefits, with an emphasis on deliberate prioritization of alternatives, although choice remains subjective, *with clearly specified assumptions, design options may be thoroughly assessed*. Contractual design thereby advances the design of maintainable software.

## 1.6 OO DESIGN PRINCIPLES

Appropriate design for class form and functionality rests on the software engineering principles of low coupling and high cohesion as do OOD principles. The **Single Responsibility Principle** (SRP) states: *Every object should have a single responsibility that must be strictly encapsulated. Thus, there is only ever one reason to modify a class.* SRP emphasizes cohesion and promotes software maintenance by focusing class functionality on a primary goal. The class designer targets use and potential reuse so type integrity is easier to preserve. The cyclicSeq design adhered to SRP: cyclicSeq encapsulated a base number, additive factor and period length in order to generate values from a cyclic arithmetic sequence. When conditions for preservation of state (implementation invariant) are consistent with expectations of use (interface invariant), the Single Responsibility Principle holds.

## 1.7 SUMMARY

In this chapter, we illustrated systematic design via the class construct, emphasizing the provision of a cohesive external interface alongside preservation of internal control. Standard class methods were differentiated by functionality (constructor, destructor, accessor, mutator, private utility, and core public utility). When relevant, we discussed language differences between C++ and C#/Java.

Programming by Contract was described as an effective means of documenting design. The clear specification of expectations across five categories (preconditions, postconditions, interface invariant, implementation, invariant, class invariant) streamlines the process of uncovering and validating design assumptions. Documentation then yields a contract between the class designer and the client. We closed the chapter by noting that the Single Responsibility design principle is sustained by the class construct and contractual design. Concepts covered here promote the development of usable and reusable software.

## 1.8 DESIGN EXERCISE

Using the concepts covered in this chapter, design a class inRange to track the number of integers queried that fall within a specified range. For example, given a range of 100 to 900, rangeObj.query(117) yields true, rangeObj.query(11) yields false, rangeObj.query(747) yields true. After these three queries, the count of integers that fell within range would be two. Your design should include internal control of state: that is, model an on/off state. Caution: this problem description is inadequate – many details are missing. For example, is the range inclusive of its boundaries, [100,900], or exclusive, (100, 900)? Can the (internal) count of queries be reset? Is the range stable? Appendix B.1 provides and analyzes a sample C# solution – a C++ solution would not be substantively different and so is not presented.

**DESIGN INSIGHTS**   *Compilers do not enforce design*

> *Contractual design identifies requirements and impediments before coding begins*

*Class Design*

Two perspectives:

External utility (client)

Internal implementation (designer)

Encapsulation minimize exposure of internal details

remove dependencies on client

client cannot put object in an invalid state

Initialization is responsibility of constructors

Accessors should not change state

Mutators control alteration of state (may reject requests)

Private utility functions support code reuse and functional decomposition

*Contractual Design:* Alternative to defensive programming

Identifies assumptions

Specifies correct usage and guaranteed response

Promotes maintainability

May reduce internal error checking

## CONCEPTUAL QUESTIONS

1. Why study OOD?

2. Identify the key differences between client and class designer.

3. Why is a constructor needed?

4. What are the benefits of private utility functions?

5. Distinguish between the three invariants of Programming by Contract.

6. How does defensive programming differ from contractual design?

7. Describe the major components of standard class design.

## CHAPTER 2

# Ownership – Abstracted but Tracked

## CHAPTER OBJECTIVES

- Outline program memory
- Identify C# and C++ differences
- Illustrate dynamic allocation
- Contrast deallocation processes
- Evaluate storage versus computation

## 2.1 THE ABSTRACTION OF MEMORY

When asked if creation of design admits constraint, American designer Charles Eames replied, "Design depends largely on constraint". Resource dependencies constrain software. Whether evident or not, memory is a resource and so its usage is a constraint. Fortunately though, most code has limited or transient use of memory. Nonetheless, software developers must be aware of programmatic memory usage, for correctness and efficiency.

Logically, memory is a physical resource that is 'owned' when it is allocated to a process. Memory is available for use or assignment when it is not allocated. Control of memory processes typically resides within the run-time environment: utilities manage the assignment (allocation) and release (deallocation) of memory while preserving data integrity. Entire books have been devoted to operating systems and resource management. Here, we provide only a brief overview of memory usage by a running program. For more information, please consult a standard operating systems text. Regardless of implementation, memory ownership should be explicitly noted and tracked.

The **CPU** accesses data stored in different physical types of memory (registers, **cache**, secondary store), all of which have different costs and performance characteristics. A cache is a smaller, faster memory store that is co-located with the processing core and holds data from main memory in order to reduce access time. Typically, *memory is viewed abstractly as a means of storage*, a uniform and unlimited resource whose low-level manipulation is left to utilities. The mapping of a relative memory address to an actual location is left to the operating system, sustaining development of portable code.

The operating system handles and allocates memory as blocks of contiguous memory cells. How a running program uses this memory does not concern the allocator. Each memory request will be filled or rejected based only on whether or not memory is available. Likewise, a running program (process) does not care how the allocator finds free blocks or reclaims memory. Processes make only the fundamental assumption that each memory block is uniquely allocated; if a request for memory is satisfied, the requestor assumes ownership.

**Virtual memory** presents storage as uniform, addressed locations without size or boundary limitations, abstracting away actual addresses of physical memory. Hardware support and operating system utilities map a virtual address to a physical address. The operating system manages memory blocks of uniform size, called pages. By overlaying (replacing) one page with another, the operating system may manipulate programs (and their associated data) that are larger than the main memory of a computer. Thus, the size of allocated virtual memory may exceed that of physical memory.

A process accesses only a few memory locations within a given timeframe, leaving most resident pages in physical memory idle. If referenced data is not currently loaded in actual memory, a page fault occurs and the needed page is swapped with a resident page. Data can be written out to secondary storage (usually a hard drive) which is much slower than main memory. Hence, system performance degrades with excessive page faults. When memory references are confined to a set of relatively contiguous blocks of memory, the number of page faults may decrease. This principle of "locality of reference" may underlie designs pursued for enhanced performance. Cache misses may also be minimized through locality of reference.

## 2.2 HEAP MEMORY

When the executable image of a program is loaded into memory, its layout can be viewed as a partition between the code section and the data section. The code section contains the software instructions while the data section holds data used and generated by the running software. The **heap** and the **run-time stack** constitute two portions of the data section. The heap consists of blocks of memory allocated for program use when explicitly requested as the program runs. Memory so allocated is called 'dynamic' because its address is not known or allocated until run-time. The size of memory so requested need not be specified until run-time.

The run-time stack stores data as it comes into scope via function calls. When a function is invoked, the **stack frame** (activation record) associated with the function is pushed onto the run-time stack, recording essential information, such as the program counter and local variables. When a function terminates, its scope is exited, and its stack frame popped off the run-time stack. A stack frame holds all variables local to the function, whether allocated by declaration, pass by value or return by value. Since the compiler assesses the size of a stack frame, its layout is determined at compile-time and, hence, is called **static allocation**.

The primary (heap) and stack allocators share the same pool of memory; each allocator starts at opposite ends of this memory chunk and 'grow' toward each other as each allocates memory. As sketched in Figure 2.1,

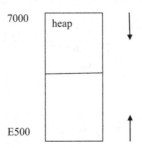

FIGURE 2.1   Program Heap and Stack Memory.

the heap allocator may start at a low address and grow upward while the stack allocator starts at a high address and grows downward. Obviously, the two must not cross. For an executing program, *heap memory is more expensive than stack allocated memory*. Why? The compiler determines the size of stack frames and, using relative addressing, lays out the required stack memory. Dedicated hardware, such as the stack register, supports fast updates to stack memory references. At run-time then, there is little overhead for processing requests for stack memory. In contrast, heap memory is controlled through the allocator (and deallocator), incurring significant run-time overhead.

A memory allocation request must specify size. In C, using `malloc` or `calloc` calls, the caller passes in the amount of memory requested using the `sizeof` operator. In C++, C# and Java, the `new` operator is invoked with a typed **pointer** or **reference**: the size of memory requested is inferred from type; an `int` may require 4 bytes of memory, a `char` 2 bytes, etc. Example 2.1 gives sample C++ allocations; Figure 2.2 illustrates the corresponding memory diagrams.

**Example 2.1  Allocating C++ (Heap) Memory at Run-Time**

```
// "ptr" is a pointer variable
// allocated on the stack
// "ptr" holds the address of the heap
// object returned by new

// #1: MyType object allocated
MyType*      ptr = new MyType;

// deallocate heap memory via
// call to delete operator
// #2: MyType object deallocated
delete ptr;

// null out pointer to indicate
// it 'points to nothing'
ptr = nullptr;              // #3: reset pointer

// pointers can also hold the address of an array
// #4: 900 MyType objects allocated
ptr = new MyType[900];
...
```

```
// must use delete[] when deallocating
// an array on heap
// #5: 900 MyType objects deallocated
delete[]  ptr;

// specify nothrow so that zero
// returned if request fails
// #6: allocation uncertain
ptr = new (nothrow) MyType[900];

if (!ptr)     cout << "Asked
  for too much!!" << endl;
// #7: conditional deallocation
else    delete[] ptr;
```

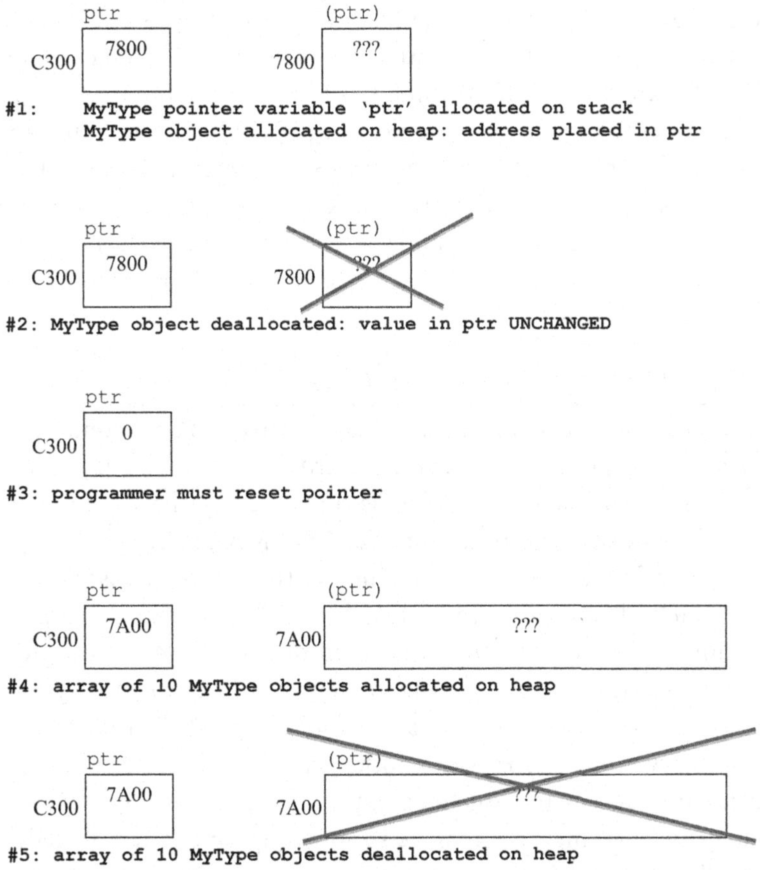

#1:    MyType pointer variable 'ptr' allocated on stack
       MyType object allocated on heap: address placed in ptr

#2: MyType object deallocated: value in ptr UNCHANGED

#3: programmer must reset pointer

#4: array of 10 MyType objects allocated on heap

#5: array of 10 MyType objects deallocated on heap

FIGURE 2.2   Heap Allocation at Run-Time.

Responding to a call to new, the allocator finds and returns the starting address of a memory block large enough to hold the requested amount of memory. The size of each block allocated is recorded so that deallocation releases the correct amount of memory. The addresses and size of memory blocks available for allocation are stored in a 'free' list; the addresses and size of allocated blocks are stored in an 'allocated' list. The allocator processes memory requests: it does not track address values held in pointer (or reference) variables. Consequently, the client MUST ensure that pointers (or references) are reset (to zero or null) when ownership of heap memory has been released!

Memory requests may fail and obviously will do so when the amount requested exceeds that available. When no single block is large enough to satisfy a memory request, even though enough memory is free on the heap, a memory request will also fail. In this case, when there is not enough contiguous memory, the heap is called **fragmented**. Consider allocating an array of 500 doubles. If a double takes 8 bytes, 4,000 bytes are needed. If 30,000 bytes are available but all free blocks are of size 3200 or smaller, the memory request will fail, and typically the allocator will throw an OutOfMemory exception. In Example 2.1, statement #4 should be wrapped in a try block if the memory request may fail. Alternatively, the call to new could specify nothrow, as in statement #6, so that the value of zero is returned, instead of a thrown exception, if the memory request fails.

## 2.3 OWNERSHIP OF HEAP OBJECTS

Stack allocation and deallocation is easy and lossless. The compiler generates and manipulates stack frames; hardware makes such processing efficient. However, stack allocation is rigid because the size of memory allocated must be known at compile time. Moreover, access to local memory in a function is transient since exiting scope releases the stack frame. Stack allocation is thus insufficient for persistent data. Heap allocation incurs the run-time overhead of calling new. Heap deallocation is complex and may be incomplete. But heap memory provides flexibility: memory requirements need not be specified until run-time, and can vary from one run to another (without code recompilation). Additionally, access to heap memory is not confined to local scope. Stack versus heap memory may be summed up as: efficient (fixed) versus flexible; localized versus persistent; secure versus vulnerable; lossless versus leaky.

**Example 2.2 Object Definition (Allocation)**

```
// C#/Java object definition:
// objects are references
//      variables zeroed out if not initialized
// #1 objA reference zeroed
myType          objA;
// #2 no-arg constructor
myType          objB = new myType();
// #3 constructor takes int
myType          objC = new myType(42);


// C++ object definition:
// by default, stack allocation
// #4 default constructor invoked
myType          objD;
// #5 constructor that takes int
myType          objE(42);


// C++ object definition: specification
// of heap allocation
// #6 objPtr1 not zeroed
myType*         objPtr1;
// #7 objPtr2 zeroed out
myType*         objPtr2 = nullptr;
// #8 call to allocator ….
myType*         objPtr3 = new myType;


// must deallocate C++ heap object
// when no longer used
// #8 heap memory released
delete          objPtr3;
```

C# and Java allocate objects only on the heap, requiring a call to the new operator for every object. C# and Java object declarations are only declarations of references that are zeroed out by default. Since the new operator is always called with a specified constructor, there are no hidden assumptions about which constructor is invoked. In Example 2.2, statement #2 invokes the no-argument constructor, statement #3 invokes an overloaded constructor that takes a passed integer value.

C++ allocates objects on the stack, by default. C++ allocates heap objects in the same manner as Java and C#, via a call to the new operator.

C++ programmers use pointers to hold the addresses of heap-allocated memory but the C++ compiler does not automatically zero out pointers upon declaration. In Example 2.2, a no-argument constructor is called to initialize objD while a constructor that takes an integer value is called to initialize objE. For stack allocated objects, it may not be evident in C++ that constructors are invoked. A C++ programmer must remember to deallocate all heap objects before their handles (the pointer variables that contain their addresses) go out of scope. Otherwise, access to the heap memory will be lost; that is, a memory leak will occur. Figure 2.3 provides memory diagrams corresponding to Example 2.2.

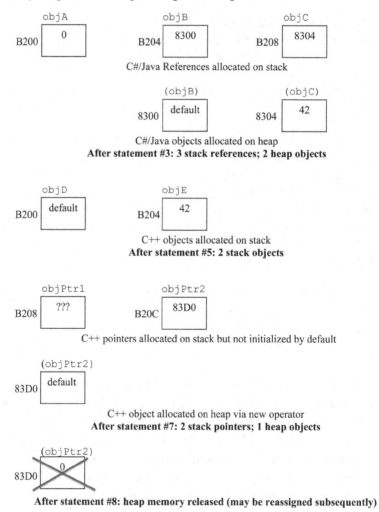

FIGURE 2.3   Memory diagrams for Example 2.2.

Caveats discussed in Appendix A apply here. The uninitialized C++ pointer, objPtr1 (statement #2), will not be zeroed out and hence contains a value – the residual bit string left in memory. If objPtr1 then is dereferenced, the residual bit string will be interpreted as a legitimate address and memory that is not owned by objPtr1 may be modified, possibly resulting in data corruption. There would be no ill effects if the invalidly referenced memory is subsequently overwritten with valid data or if execution terminates before this memory is accessed again. A run-time exception is also possible if the interpreted address is reserved for the operating system. Without proper initialization, it is hard to predict behavior. C++ design guidelines stress the need to zero out pointers upon declaration and when memory is released so that a non-zero value may be interpreted as a legal address.

## 2.3.1  Array Allocation

Array elements must be allocated contiguously. Consequently, requests for large arrays are more likely to fail (the new operator throws an exception) due to insufficient or fragmented heap memory. Unless specified, a failed memory request does not return a zero: programmers must specify a 'nothrow' form of the new operator so that zero is returned to indicate a failed memory allocation request. Alternatively, exceptions can be used to avoid run-time errors. For details on secure coding, see [Sea13].

**Example 2.3  Array Allocation: C# vs C++**

```
// C# (and Java) object arrays: array of references
// #1  an array of 100 references allocated
myType[]     db = new myType[100];

// #2  each C# reference individually initialized
//     to hold address of heap-allocated object
for (int j = 0; j < db.Length; j++)
        db[j] = new myType(j);

// #3  C++: array of objects allocated on stack
//      default constructor implicitly
//      invoked for each object
// PROBLEM IF CLASS myType DOES NOT
// HAVE NO-ARGUMENT CONSTRUCTOR
myType        db[100];
```

```
// #4      may overwrite default C++ initialization
for (int j = 0; j < 100; j++)
{    myType local(j);   // non-default constructor
     db[j] = local;    // #4.2
}


// #5   C++: if myType has only
// constructors with arguments
//      => cannot use array of objects
//      => HEAP MEMORY requires
//         TRACKING OWNERSHIP (deallocation)
//      => use STL container OR array of pointers
myType*   dbP[100];
for (int j = 0; j < 100; j++)
        dbP[j] = new myType(j);
```

Allocating an array of objects requires multiple steps in C# and Java. First, a reference to an array is declared and then initialized to hold the address of an array of references. These two actions may be coded as separate statements but are commonly combined, as in statement #1 of Example 2.3. Finally, element by element, the array of references is initialized to hold addresses of individually allocated objects, as shown in statement #2. C# is not dependent on the no-argument constructor since a (any) constructor must be explicitly identified when calling new. See Figure 2.4.

### 2.3.2 Design Intervention

Details may be tricky for C++ object arrays because each element of the array is an object that, like other C++ objects, is implicitly initialized by a

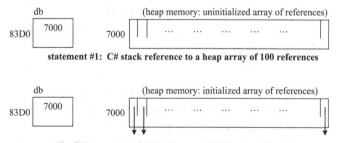

FIGURE 2.4    Memory diagrams for Example 2.3.

constructor upon declaration. The C++ array declaration of statement #3 in Example 2.3 requires that the compiler call the no-argument constructor; syntactically, another constructor cannot be specified. The default initialization may subsequently be overwritten, as shown in statement #4.2 (as well as in Example 1.3). If a class does not provide a public no-argument constructor, an array of C++ objects cannot be allocated. Typical workarounds are to use: 1) a STL container rather than an array; or 2) an array of pointers to hold addresses of objects individually allocated on the heap, as shown in statement #5, just as is done in Java and C#. Both solutions may impede performance requirements. Additionally, the second option requires that the client track ownership of heap memory and ensure object deallocation.

Usually, a C++ class designer should provide a no-argument constructor to support the allocation of object arrays. By design, the no-argument constructor could put the object in an unusable state, forcing an overwrite that utilizes a constructor with parameters. Example 2.4 rewrites the cyclicSeq class from Example 1.7 in C++ to demonstrate such a design: the no-argument constructor puts the object in an 'off' state. An array of C++ cyclicSeq objects may be allocated but the client cannot effectively use the array until each element is overwritten. Contractual design must specify the constraints of the no-argument constructor, stating support for array allocation alongside a disclaimer that overwriting is expected.

**Example 2.4 Design to Facilitate C++ Array Allocation**

```
class cyclicSeq // a cyclic arithmetic generator
{       bool        on = true;
        unsigned    place = 0;
        unsigned    period;
        unsigned    a1;
        unsigned    dst;
    public:
      cyclicSeq(unsigned start,unsigned inc,
      unsigned lgth = 100)
      {     a1 = start;
            dst = inc;
            period = lgth;
      }

      // no-argument constructor sets
      // object in unusable state
```

```
        cyclicSeq()
        {       a1 = dst = period = place = 0;
                on = false;
        }
        int nextNum()
        {       if (!on)     return -1;
                place = (place + 1)% period;
                return int(a1 + dst*place);
        }

        bool expand(unsigned scale)
        {       if (!on || scale == 0)       return false;
                period *= scale;
                return true;
        }

        bool isOn()       {       return on;       }

        bool toggleOn()
        {       if (!on && a1==0 && dst==0
                && period==0 && place== 0)
                    return false;
                return on = !on;
        }
};
```

### 2.3.3 Persistent Data

Stack allocated memory is released upon exiting scope. Heap memory is used for persistent data. Ownership of heap memory is often passed out of a function, making it unlikely that a delete is executed in the same scope as its corresponding new. C++ programmers must track and deallocate heap memory. Yet, matching every new to a delete is difficult to guarantee across function calls and aliases, and in the event of a thrown exception.

The first function in Example 2.5 calls the new operator to allocate a single integer on the heap. When the call executes at run-time, the address of the heap integer so allocated is returned and assigned to the local pointer variable heapInt. The second function in Example 2.5 calls the new[] operator to allocate an array of 100 integers on the heap. When the call executes at run-time, the address of the first element of the array of 100 heap integers is returned and assigned to the pointer variable heapIntA.

### Example 2.5  C++ Memory Management

```
// C++ code: allocation, transfer
// and deallocation of heap memory
void   matchNewDelete1()
{        int*  heapInt = new int;
         ...
         delete  heapInt;
}


void   matchNewDelete2()
{        int*  heapIntA = new int[100];
         ...
         delete[] heapIntA;
}


int*  transferOwnershipOut(int threshold)
{        int*  heapInt = new int;
         *heapInt = threshold;
         ...
         return  heapInt;
}


void assumeOwnership(int*& ptrPassedByRef)
{        int*  heapInt = ptrPassedByRef;
         // caller no longer owner
         ptrPassedByRef = 0;
         *heapInt = 999;
         ...
         delete                heapInt;
}


int*  assumeThenTransfer(int*&  ptrPassedByRef)
{        int*  heapInt = ptrPassedByRef;
         // caller no longer owner
         ptrPassedByRef = 0;
         ...
         return  heapInt;
}
```

```
....
matchNewDelete1();                              // call #1
matchNewDelete2();                              // call #2

int*  myPtr = transferOwnershipOut(33); // call #3
*myPtr = 21;

assumeOwnership(myPtr);                          // call #4
if (myPtr)    cout << *myPtr << endl;    // no output

myPtr = transferOwnershipOut(55);               // call #5
// call #6
int*    yourPtr = assumeThenTransfer(myPtr);
if (myPtr)      cout << *myPtr << endl;   // no output
// 55 output
if (yourPtr)    cout << *yourPtr << endl;
```

In C# and Java, explicit allocation, via new and new[], is comparable to C++ but is not associated with responsibility for deallocation. Java and C# provide implicit deallocation via garbage collection which reclaims allocated but inactive memory from the heap. Garbage collection frees the programmer from the headaches of tracking memory but has its own drawbacks: 1) program execution must be suspended for the garbage collector to run; 2) garbage collection is not perfect (not all garbage is marked for reclamation); and, 3) before collection, performance is degraded by **heap fragmentation** (the scattering of unused blocks among free blocks which reduces the size of contiguous memory). To decrease the amount of memory allocated but left lingering on the heap until the garbage collector runs, design should: 1) minimize the use of temporaries; 2) transfer ownership; 3) share memory.

C++ requires explicit deallocation via the delete operator. Released heap memory may be subsequently reassigned. For arrays, the delete[] operator should be invoked. In C++, every new should be matched with a delete, and every new[] should be matched with a delete[]. Aliases, transfer of ownership, parameter passing, exceptions, etc., make this simplistic design guideline difficult to follow.

Example 2.5 illustrates appropriate management of heap memory. Functions isolate memory allocation and clearly identify when memory is released or the responsibility to do so (ownership) is transferred. Figure 2.5 provides the corresponding memory diagrams. matchNewDelete1() and matchNewDelete2() illustrate the allocation and deallocation of heap memory in the same scope. It is easy then to verify that no memory leaks as long as there is no premature exit (as when an exception is thrown) in between allocation and deallocation.

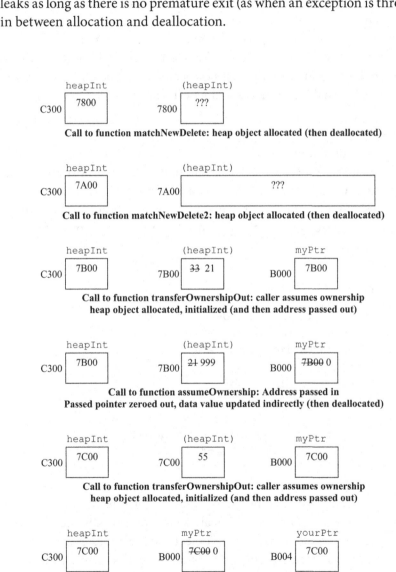

Call to function matchNewDelete: heap object allocated (then deallocated)

Call to function matchNewDelete2: heap object allocated (then deallocated)

Call to function transferOwnershipOut: caller assumes ownership
heap object allocated, initialized (and then address passed out)

Call to function assumeOwnership: Address passed in
Passed pointer zeroed out, data value updated indirectly (then deallocated)

Call to function transferOwnershipOut: caller assumes ownership
heap object allocated, initialized (and then address passed out)

Call to function assumeThenTransfer: Address passed in
Passed pointer zeroed out (and then address passed out)

FIGURE 2.5  Memory diagrams for Example 2.5.

`assumeOwnership()` and `assumeThenTransfer()` assume ownership of heap memory passed in by the caller (via a pointer passed by reference). The `int` pointer passed by reference into each function is zeroed out so that the caller can no longer access the transferred memory. Upon return from either function call, the parameter `ptrPassed ByRef` points to nothing, consistent with the caller's release of ownership. By zeroing out the pointer passed by reference, each function records its assumption of ownership. It is crucial to track ownership. If ownership is transferred, the pointer that releases ownership must be nulled or zeroed out, indicating that it does not contain a valid address.

## 2.4 CLASS DESIGN

In C++, class design *must* consider ownership when heap memory is allocated internal to an object. In C#, class design should consider ownership even though C# does not explicitly deallocate heap memory. In any language, design may improve performance (and the efficacy of the garbage collector) by consciously tracking ownership and zeroing out references when appropriate.

### Example 2.6 C++ Unseen Leaks

```
// client code uses hiddenLeak and noLeak objects
//    NO CALLS to NEW => calls
//    to DELETE inappropriate
void strangeFn()
{        hiddenLeak      objA;
         noLeak          objB;

         cout << "I am following design
                 guidelines" << endl;
}
```

The type names of Example 2.6 suggest that one object leaks but the other does not. For stack allocated objects, this distinction does not appear to make sense. Both objects are allocated memory in `strangeFn()`'s stack frame, which is popped off the run-time stack upon function exit. Where is the memory leak? Not on the stack. Is there a memory leak on the heap? There is no explicit call to the new operator in Example 2.6. The client code is correct. However, the internal structure of neither type of object is known. Perhaps the class design is not correct. The `hiddenLeak` constructor in Example 2.7 allocates heap memory via a call to the `new` operator. But the class code does not contain a matching `delete`: there is no destructor to deallocate heap memory. Hence, the leak. A destructor should be defined in every C++ class that allocates heap memory. Output statements, though normally inappropriate in a class definition, have been added to the constructors and destructor of Example 2.7 to identify this discrepancy.

**Example 2.7  C++ Class Design Must Address Memory**

```
// IMPROPERLY DESIGNED: heap memory
// allocated in constructor
//      no destructor, no deallocation:
//      no delete[] to match new[]
class hiddenLeak{
   private:
         int*    heapData;
         int     size;
   public:
         hiddenLeak(unsigned s = 100)
         {   size = s;
             heapData = new int[size];
             cout << size << "  ints allocated
             hiddenLeak" << endl;
         }
};

// Heap memory: constructor allocates;
// destructor deallocates
class noLeak{
   private: // need copy semantics - see Chapter 3
         int*    heapData;
         int     size;
   public:
         noLeak(unsigned s = 100)
         {   size = s;
             heapData = new int[size];
             cout << size << "  ints allocated
             noLeak" << endl;
         }
   // destructor deallocates heapData
   ~noLeak()
```

```
{     if (heapData)          delete[] heapData;
cout << size << "  ints DEallocated noLeak" << endl;
}
};
```

The C++ compiler automatically patches in a call to the destructor when a stack object goes out of scope, or when a heap object is deallocated via the delete (or delete[]) operator; the client should not directly invoke the destructor. Essentially, the destructor is a cleanup routine: it performs any actions, such as deallocating heap memory that must be executed before an object goes out of scope. Although some C++ design guidelines suggest that class designers should always define a destructor, many class designs meet expectations without a destructor. When is a destructor required? Minimally, when an object allocates or assumes ownership of heap memory. The class designs in Example 2.7 appear similar, except that the noLeak class defines a destructor which provides a delete to match the new call in the constructor. We examine destructors relative to design longevity in Chapter 7. Safe and effective memory management is guaranteed through **smart pointers** which are examined in Chapter 3.

When the (stack) objects go out of scope in Example 2.6, the noLeak destructor is implicitly invoked and the heap memory allocated internal to objB will be deallocated. There is no hiddenLeak destructor to invoke. Hence, the heap memory allocated internal to objA remains allocated but is unused: the pointer (handle) that provides access to that memory, objA.heapData, goes out of scope when objA goes out of scope. Memory leaks can be prevented with destructors. **C++ class design also requires explicit decisions with respect to copying**, which is examined in Chapter 3.

## 2.5 MEMORY RECLAMATION

Correct memory management rapidly becomes complex: aliases, transfer of ownership, call by reference, exceptions and shared use all complicate the task of tracking ownership. *It is challenging to reliably ensure deallocation of heap memory.* Failure to release memory yields a memory leak. Without aliasing, once a handle goes out of scope, the memory so referenced is no longer accessible: the memory remains allocated on the heap (its block address remains on the allocated list) even though it can no longer be accessed. The allocator does not know that this memory is inaccessible. To the allocator, the block is still in use because no release request was received.

## 2.5.1 C++ Explicit Deallocation

In C and C++, deallocation is explicit: the programmer must call `free` in C or `delete` (or `delete[]` for arrays) in C++ to release memory allocated on the heap. The freed memory block is taken off the allocated list and returned to the free list. Normally, there is no output in response to the explicit release of a memory block. However, if an invalid pointer is supplied as the base address of the block, an exception may be thrown. A common mantra for C++ programmers is "match every `new` with a `delete`". If only design could be so simple! Due to scope, aliasing, and transfer of ownership, matching each `new` with a `delete` is not a trivial endeavor. Modern C++ provides smart pointers as a safe alternative to **raw pointers** (the standard pointer construct that does not provide built-in resource management). Java and C# do not use explicit deallocation.

## 2.5.2 Garbage Collection

Memory reclamation may be explicit (calls to `delete` or `free`) or **implicit deallocation (garbage collection)**. As a background process, the garbage collector is invoked when available heap memory is insufficient or too fragmented. Garbage collection tags all allocated blocks that are reachable, reclaiming blocks that are not reachable (and thus unusable). Data is reachable if it is directly addressed by a currently active variable, or indirectly addressed via a pointer (reference) embedded in an active variable (or in a variable referenced by an active variable, etc.). The time between when data becomes inactive and when it is reclaimed may be significant. The legitimacy of a client request for reclamation is questionable. In Java, the call `System.gc()` is only a suggestion, not a directive. In C#, the call `GC.Collect()` may impede performance, especially if used when not warranted. Without explicit deallocation, memory tends to stay allocated so the heap may be more prone to fragmentation under garbage collection.

Software design may strive to minimize memory usage since memory access, allocation and reclamation impact performance. Heap allocation finds an appropriately sized free block, marks it allocated, and returns its base address. The overhead of allocation is unavoidable and depends mostly on the first step – the search for available memory in the heap. The *overhead of deallocation varies.* There is no reclamation overhead if garbage collector does not run. With implicit deallocation though, the heap is more likely to become fragmented, resulting in a costly search for free blocks during allocation and, thus, degraded performance. The *overhead of explicit deallocation is borne incrementally*, as each deallocation request is processed.

Garbage collection is the 'automatic' reclamation of heap memory no longer in use (garbage) and, thus, removes the responsibility of memory deallocation from software developers. Small applications that do not use much memory, or applications that run for only a limited time before termination (or restart), may never require garbage collection. *Garbage collection incurs no overhead unless the garbage collector runs.* Garbage collection strategies date back decades to the development of the programming language LISP. The popularity of Java renewed interest in garbage collection algorithms and analyses. Research continues because garbage collection is not a perfect process: the identification of garbage must be conservative and therefore is incomplete. Memory leaks still exist in Java and C#. For advanced readings and current research, please consult sun.com.

**Example 2.8 Classic Mark and Sweep Algorithm for Identifying Garbage**

```
// start with direct references, the root set:
//      all visible variables (active memory)
//      at time of sweep
// trace out to all variable indirectly referenced
void markSweep()
{       for each Object r in rootSet
              mark(r);
}


// recursive depth-first marking
// terminates when all reachable objects marked
void mark(Object x)
{       if (!x.marked)
        {       x.marked = true;
                for each Object y referenced by x
                      mark(y);
        }
}
```

```
// if heap object marked: KEEP
//      clear marked status in preparation
//      for subsequent sweeps
// if heap object unmarked: RECLAIM (garbage)
void sweep()
{
        for each Object x on heap
                if (x.marked)          x.marked = false;
                else                   release(x);
}
```

An executing program is suspended when the garbage collector runs. Suspended processing may not be a viable option for many real-time applications but may not be noticed in others. The stack and static memory of the suspended process yields a **root set** of active variables (data). All memory blocks associated with this root set are marked and recursively examined for embedded references to other data. Example 2.8 illustrates the classic mark & sweep algorithm for marking all reachable data. Each recursive level of mark(y) corresponds to another step in a chain of references traced from the root set. Recursion terminates when no additional blocks are marked, indicating that all reachable memory has been marked. sweep()then sweeps through the heap, reclaiming all unmarked blocks of memory.

Garbage collection distinguishes between live and dead data. When data is no longer used, it should be considered 'garbage' and reclaimed. However, only inaccessible objects (data variables) are collected. All variables reachable from the root set are marked for preservation even if such variables are not actively used and should be reclaimed. *A reference that holds an address of an unused object prevents that object from being reclaimed.* The mark and sweep algorithm will not reclaim any active blocks but may fail to reclaim all inactive blocks. An object that is no longer used but is still accessible because its heap address 'lingers' in some reference cannot be reclaimed. Garbage collection may, in fact, miss some garbage. Design guidelines recommend the nulling or zeroing out of pointers and references once an object is considered inactive.

Since garbage collection is not a perfect process, C#/Java professionals should track memory ownership. Reinterpreting the responsibility to deallocate/delete, as the nulling or zeroing out of a reference, would recast C++ guidelines for Java and C#. Why? Effective garbage collection depends on appropriate values in references: that is, nonzero for valid addresses and zero for inactive references. C# and Java provide weak references to assist garbage collection. For details, see msdn.microsoft.com or sun.com.

### 2.5.3 Reference Counting

**Reference counting** explicitly tracks aliases and disperses the cost of memory management across all allocation and deallocation requests. Each memory block has a counter associated with it that indicates how many references (or handles) refer to that memory block. If the reference count is zero then the memory block may be reclaimed since it is no longer in use. Reference counting cannot detect cyclic references and thus may not collect all garbage. Figure 2.6 displays a cycle of four objects (allocated data blocks). There are no external references to this cycle. Yet, none of the four blocks will be reclaimed because each has a positive reference count, due to a reference from another (unused) block in the cycle.

Large data collections are more efficiently and securely managed when data is shared rather than copied. shared _ ptr provides the means to do so, as examined in the next chapter. At a design level though, to avoid costly copying, reference counting may be implemented on a class-level, mimicking its implementation as a utility. A class may centralize data access, providing a public, static instantiation routine. Upon the first request to instantiate, data is allocated and the reference count set to one. Subsequent requests to instantiate involve no allocation of memory, simply an increment of the reference count: the same address is returned for all instantiation requests. Aliases abound in this scheme. Deallocation requests decrement the reference count. When the reference count reaches zero, the data may be deallocated.

Unlike reference counting, mark and sweep algorithms are not often implemented in customized software. We examined this approach to illustrate the effect of retaining a reference to memory that is no longer used: the memory block cannot be reclaimed. Specious references fragment the heap and degrade performance. The explosive growth of Java led to its quick adoption for many software projects. Many programs were designed without adequate consideration of memory usage, leading to performance

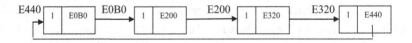

FIGURE 2.6   Reference Counting defeated by Cyclic References.

degradation, and poor scaling since expanded demand for memory could not be satisfied. **Profilers**, software that measures time complexity and memory usage of code, followed Java on the market as developers sought to uncover heap fragmentation and sources of inefficiency.

Neither reference counting nor mark-and-sweep algorithms tackle heap fragmentation, other than incidentally by returning blocks that can be coalesced with adjacent free blocks. Heap **compaction** is a separate process but may follow (or be interwoven with) garbage collection. Compaction reassigns heap memory still in use to one end of the heap in order to maximize the amount of contiguous memory available for future allocation, and to minimize fragmentation. Compaction is expensive and complicated. *To constrain heap fragmentation, design recommendations include minimizing both allocations and the use of temporaries.*

Memory management strives to ensure data validity, consistent performance, and correct deallocation. However, doing so is difficult, see Table 2.1. Data corruption occurs when hidden aliases permit uncoordinated updates to memory (see Chapter 3). *Errors due to data corruption are difficult to detect because they often occur far from their source.* Performance degradation is often traced to a fragmented heap. Memory leaks occur when heap memory is not released before handles goes out of scope.

C++ leads the growth of large-scale software development, with a concomitant increase in the number of software developers, many without extensive knowledge of software design or experience with machine hardware. Hence, the reliability of programmer-managed memory became a

TABLE 2.1   Common Difficulties with Program Memory

|  | Condition | Cause | Effect |
|---|---|---|---|
| **Data Corruption** | Memory overwritten | Ownership undermined Hidden aliases | Undefined behavior |
| **Performance Degradation** | Allocator overhead to find free memory | Fragmented heap | Poor Scalability |
| **Memory Leak** | Heap memory not released | `delete` not executed Garbage not identified | Diminished resource |
| **C++ Memory Leak** | Memory inaccessible | Lost handle to memory | Diminished resource |

significant concern. Experienced professionals developed coding principles to reduce memory errors and overhead, including: 1) match every new with a `delete`; 2) transfer ownership; 3) track aliases; 4) define classes with constructor(s), destructor and copy semantics (see Chapter 3). However, *design guidelines cannot be enforced and thus are always insufficient.* The popularity of garbage collected languages such as Java, and C# is somewhat due to reduced memory management responsibilities.

## 2.6 DESIGN: STORAGE VERSUS COMPUTATION

Memory management is complex but important for correctness and performance. Although low-level memory management details can and should be ignored, software design may need to consider memory. Memory access can become the bottleneck of a system. Processor speeds cannot rescue a data-intensive system from a poor design. In general, one trades space for performance. Increased storage requirements are justified when, say, calculations are stored to avoid repeated computation. If specific computations are frequently made, and if the data values involved in such computations are infrequently modified, then storing computations for future lookup is a reasonable design option.

For example, given a stable data set that must support frequent queries that are dependent on the mean value of the set, a design that stores the mean likely yields better performance than a design that calculates the mean upon each request. Conversely, given an unstable data set (where values are frequently inserted or deleted) that must support infrequent queries, a design that calculates the mean upon request may reduce computational overhead. Example 2.9 contrasts these two different designs.

**Example 2.9  Storage versus Computation**

```
public class storeMean
{       private List<int> values
                          = new List<int>();
        private float           mean;

        public  void add(int x)
        {       values.Add(x);

                int sum = 0;
                foreach (int k in values)
                        sum += k;
```

```
            mean = (float)(sum / values.Count);
    }

    public  void delete(int x)
    {       values.Remove(x);

            int sum = 0;
            foreach (int k in values)
                    sum += k;
            mean = (float)(sum / values.Count);
    }

    public float Mean() {  get => mean; }
}

public class computeMean
{       private List<int> values
                        = new List<int>();

    public  void add(int x)
    {       values.Add(x);          }
    public  void delete(int x)
    {       values.Remove(x);       }

    public float getMean()
    {       int sum = 0;
            foreach (int k in values)
                    sum += k;
            return (float)(sum / values.Count);
    }
}
```

Design may depend on estimates of query frequency and data stability. The more stable a data set, and the more frequent inquiries, the greater the benefit of storing values like the mean (minimum, maximum, etc.). The converse also holds: frequent changes in data values alongside infrequent queries suggest that computation upon demand is more efficient. Caution is recommended when a class design uses unbounded storage, such as the generic list type. Why? Generic containers are so easy to use that developers may not notice how much is being stored, whether such storage is necessary, and the costs of resizing, reordering, etc.

Consider the problem of monitoring access to data. The idea is to encapsulate a number and process queries that test whether a proffered number is less than, equal to, or greater than the internal number. Also, track the number of queries processed that were less than, equal to, or greater than the tracked number. For example, if a query passed in '35' to an object that encapsulated '43' then the 'lessThan' count would be incremented. Example 2.10 illustrates a simple design.

**Example 2.10  Counts Stored; Queries Discarded**

```
public class   tracker
{       private int   num;
        private int   less = 0;
        private int   equal = 0;
        private int   more = 0;

        public tracker(int x) {   num = x;                }

        public int query(int y)
        {       if (y < num)
                {       less++;        return -1;          }
                if (y == num)
                {       equal++;       return 0;           }
                more++;
                return 1;
        }

        public int getLessCt()
        {       return less;    }

        public int getEqualCt()
        {       return equal;   }

        public int getMoreCt()
        {       return more;    }
}
```

An overdesigned response is presented in Example 2.11. A list stores each query. Upon demand, the entire list is scanned to determine the relative counts of numbers that fell below, matched, or were above the encapsulated value. A tremendous amount of storage is used, <u>and</u> computational overhead is increased. *Reassess any design that maximizes both storage and computation.*

**Example 2.11 Queries Stored**

```
public class  tooBig
{        private int               num;
         private List<int>
         queries = new List<int>();

         public tooBig(int x) {        num = x;        }
         public int query(int y)
         {        queries.Add(y);
                  if (y < num)              return -1;
                  if (y == num)             return 0;
                  return 1;
         }

         public int getLessCt()
         {        int count = 0;
                  foreach (int q in queries)
                           if ( q < num) count++;
                  return count;
         }

         public int getEqualCt()
         {        int count = 0;
                  foreach (int q in queries)
                           if ( q == num) count++;
                  return count;
         }

         public int getMoreCt()
         {        int count = 0;
                  foreach (int q in queries)
                           if ( q > num) count++;
                  return count;
         }
}
```

## 2.7 OO DESIGN PRINCIPLES

Memory must be managed correctly by both the client and the class designer. As much as possible, responsibility should be internalized. Memory management is more difficult in C++ than in C#. C++ relies on explicit deallocation and does not zero initialize data (due to prioritization of efficiency). Memory access errors are more likely when a variable is not assigned a valid, initial value. In contrast, C# zero initializes all data. Yet, any class designer can establish safe and effective memory use and initialization. The responsibility driven design principle generalizes this idea of identifying internal requirements and state control.

### 2.7.1 Responsibility Driven Design Principle

*Identify all object responsibilities (functionality) and required information*

Responsibility-driven design works in tandem with contractual design to specify assumptions and expectations as to use. The implementation invariant specifies the design of the object, as well as functionality and internal responsibility. The interface invariant specifies any client responsibility for consistent use. A clear example of a responsible design is a priority queue type that internally resizes to expand capacity when needed and periodically ages stored data items so as to prevent starvation. The class invariant would specify unbounded capacity, prioritization and aging of items (aging is akin to modifying priority) and defined error response. The implementation invariant would record design decisions for storage and aging.

## 2.8 SUMMARY

Modern programming languages abstract away most memory management details. Software developers can design portable and maintainable code when it is not directly tied to specific memory addresses. To illustrate differences in program memory management, we contrasted allocation and deallocation in C++ and C#. See Chapter 3 for detailed analyses of copying. Memory management is not a trivial endeavor; no approach easily prevents all memory leaks and data corruption. Hence, the competent developer should understand the memory models of different programming languages and their effects on software design.

At the chapter end, we contrasted different designs: data storage for lookup (to reduce computational overhead) versus computation upon demand. Design evaluation must explicitly consider tradeoffs: memory requirements; frequency of computation requests; data stability. Extra memory can be justified by enhanced performance. Programmers should remember though that computation is typically fast while memory access remains slow.

## 2.9 DESIGN EXERCISE

To apply the concepts covered in this chapter, define a `feeLedger` class, essentially a container to track fees, identifying the minimum, maximum, mean and median values. Capacity must be unbounded. Do not consider copy semantics – the details of copying within a class design are covered in the next chapter. Appendix B.2 provides and analyzes a sample C++ solution.

## DESIGN INSIGHTS

*Memory*

Viewed abstractly and thus treated uniformly

Cost of access dependent on location (cache, secondary store, etc.)

Heap memory provides flexibility and persistence

*but is vulnerable and leaky*

Heap memory is more expensive than run-time stack

Compiler lays out stack frames, no run-time overhead

Heap memory allocated at run-time via call to allocator

Heap fragmentation dampens performance, *to reduce fragmentation*

Minimize allocations and use of temporaries

Language Differences for managing program memory

C++: explicit deallocation of heap memory

C#/Java: implicit deallocation of heap memory (garbage collection)

*Software Design*

C++ programmer must manage memory

Memory leaks prevented when 'every `new` matched with a `delete`'

Memory must be deallocated before (last) handle goes out of scope

Matching 'every new with a `delete`' difficult

Parameter passing, transfer of ownership, aliasing

Objects encapsulate dynamic memory allocation

→ obscured need for direct memory management

Design guidelines not enforced by compiler, and, thus, are inadequate

## CONCEPTUAL QUESTIONS

1. What are the advantages and disadvantages of heap memory?

2. How does implicit and explicit deallocation differ?

3. Why is tracking the ownership of (heap) memory difficult?

4. Why is garbage collection imperfect?

5. Why is reference counting imperfect?

6. When is storage preferred over computation?

## CHAPTER 3

# Data Integrity

## CHAPTER OBJECTIVES

- Identify causes of data corruption

- Design for Data Integrity

- Examine copying design options

  - shallow, deep and suppressed copying

  - relevant C# and C++ differences

## 3.1 DATA CORRUPTION

A classic saying in software development "fast, good, cheap – choose any two" became known as the Triple Constraint or the Iron Triangle. Used in project management, this directive confronts the difficulty of optimizing speed, quality, and cost simultaneously. When it comes to data integrity though, must we choose? Stack allocated data implicitly achieves all three qualities. Here, we examine how to efficiently preserve the integrity of heap allocated data while minimizing client responsibility. Language differences arise and so are noted.

Since C# (and Java) object declarations are merely references, a call to the new operator is required to allocate an object. Thus, all C# objects reside on the heap, automatically providing data persistence. C# zeroes out declared but uninitialized variables, preventing invalid memory access due to invalid addresses. However, passing objects (references) by value in C# and Java is not secure because address transmission yields **aliases** and potentially uncontrolled state change.

Example 3.1 displays a C# class, aboveMin, with a public method that conditionally alters an encapsulated int, alongside sample client code that passes an aboveMin object by value. Figure 3.1 traces corresponding memory modifications. Pass-by-value provides separate memory for the formal parameter p (in function insecureFn) to be initialized with the value of the actual argument. Changes to the formal parameter within

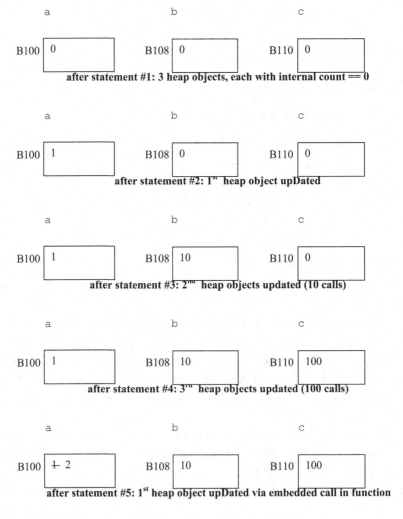

FIGURE 3.1   Three aboveMin objects.

the function then do not affect the actual argument. Yet, after the function call in statement #5, object a has an internal count of 2 rather than 1. Why? Both formal parameter p and actual argument a hold the same value – B100 – and thus address the same aboveMin heap object. Thus, p.upDate() modifies the memory of the object at location B100, that is, object a even though a was passed by value.

**Example 3.1  C# Pass by Value (Objects Are References)**

```
public class aboveMin

{        private bool active = true;
         private int   count;
         private int   min;

         public aboveMin(int v)      { min = v; }

         public bool upDate()
         {   if (!active) return false;
             count++;
             return count > min;
         }

         public int Min        { get  => min; }
         public int Count      { get  => count; }
}
// object (reference) passed by value
//      => address copied into local reference (p)
//      => both p and actual argument reference same
//      heap memory
void insecureFn(aboveMin p)
{       p.upDate();
        return;
}

aboveMin a = new aboveMin(5);
aboveMin b = new aboveMin(10);
aboveMin c = new aboveMin(15); // #1: 3 heap objects

a.upDate();   // #2 equivalent to aboveMin.upDate(B100)
              // => a.count == 1
for (int k = 0; k < 10; k++)
```

```
// #3 equivalent to aboveMin.upDate(B108)
     b.upDate();
                    // => b.count == 10
for (int k = 0; k < 100; k++)
// #4 equivalent to aboveMin.upDate(B110)
     c.upDate();
                    // => c.count == 100
// #5 pass by value not secure for objects
insecureFn(a);
                // => a.count == 2
```

Pass-by-value appears secure. Any change to a formal parameter in a function will not alter the actual argument because pass-by-value copies the actual argument to local storage in the function's stack frame. Since C# objects are references though, the formal parameter has the same address value as the actual argument and thus references the same heap object. The invocation of state-changing method upDate() through formal parameter p thus 'invisibly' alters actual argument a. In any language, pass-by-value is not secure when the argument passed is an address.

## 3.2 COPYING

Rewritten in C++, Example 3.1 would not expose objects to change via pass-by-value if the actual argument passed was a stack-allocated object. The C++ compiler automatically provides a default **copy constructor**, invoked upon pass-by-value, to copy the values from the original object (the actual argument) into memory allocated for the new object (the formal parameter). The same construction of a new object from an existing object occurs with return by value. However, this default copy constructor is insufficient for heap allocated objects. For persistence, C++ programmers must allocate data on the heap, via the new operator, yielding the same indirect manipulation of objects (via addresses) as C#, and the same potential for aliasing. Copying is more complicated with objects, aliases, and heap memory than with built-in types.

### 3.2.1 Shallow versus Deep Copying

Many developers do not think about copying because it occurs 'automatically'. When copying is needed for assignment, or for call by value, the compiler generates code for a bitwise copy. The bit string that resides in the source (the right-hand side of the assignment statement, or the actual

Two objects with different memory references (internal pointers)

FIGURE 3.2    Shallow Copy: objA = objB.

argument in a function call) is copied into the memory of the destination (the left-hand side of assignment, or the formal parameter, respectively). For primitives (which are allocated on the stack), this form of copying, called **shallow copying,** works well. However, it produces aliases when the values copied are addresses, as shown in Figure 3.2.

Figure 3.2 shows the result of a bitwise copy for the assignment objA = objB. Access to heap memory located at 8104 is lost because the pointer encapsulated in objA is overwritten with the address value from objB (8504). Now both objects address the same memory so objA can change data that objB assumes that it still owns. Shallow copying may yield data corruption and, in C++, a memory leak (access to 8104 is lost but the memory remains allocated).

When an object encapsulates references or pointers to address heap memory, should the assignment objA = objB yield aliasing or trigger a true copy? Shallow copying provides the former. To initialize a replica, additional memory is allocated and the values in the source memory are copied into the newly allocated memory. Allocation of heap memory associated with objA is identical in size and value to the separate heap memory associated with objB. This process, called **deep copying,** yields a true copy because values that are addresses are not copied directly. Figure 3.3 illustrates the memory layout of two distinct objects, where each object has two data members: a handle (pointer or reference) to heap memory, and an integer value (2 and 200, for this example).

### 3.2.2 C++ Copying of Internal Heap Memory

The new operator returns the address of allocated heap memory to its caller. The common expectation is that the caller retains ownership until

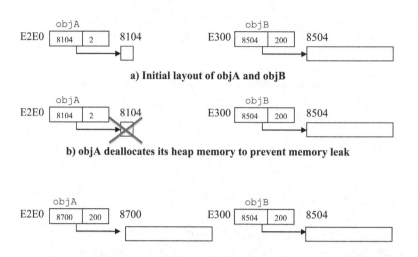

a) Initial layout of objA and objB

b) objA deallocates its heap memory to prevent memory leak

c) objA gets new heap memory, of appropriate size, and copies values from objB

FIGURE 3.3    Deep Copy: `objA` = `objB`.

release (via `delete` operator or by transferring ownership). An unseen alias may give the impression of ownership rather than shared memory, possibly leading to data corruption and, in C++, premature deallocation. Copying addresses establishes aliases. Although aliases support the sharing of data and may reduce data redundancy and inconsistency, aliases are problematic when not tracked. If two different pointers (references) hold the same address to a heap object, as in Figure 3.2, and each pointer assumes that it owns this heap memory, then data values can be changed through either pointer (or, in C++, deallocated) without regard to any other 'owner'.

In C++, a class with a properly defined destructor, but without proper copy semantics, unwittingly permits data corruption. For example, if `objB` goes out of scope before `objA`, triggering invocation of its destructor and deallocation of memory at 8504, then `objA` would point to memory that it no longer owns. Released heap memory could be reassigned by the allocator. If so, `objA` would point to memory 'owned' by another object, possibly yielding data corruption because both objects assume exclusive access. Premature deallocation, data corruption, and memory leaks are all undesirable side-effects of poorly managed aliasing.

Figure 3.3 illustrates deep copying: 1) the value of every non-pointer field is copied directly from one object to another; 2) additional heap memory is allocated and then the data values indirectly addressed via the source pointer are copied. After a deep copy, the source and destination objects have the same sized heap memory, with the same data values, but they do not share the same memory space. Subsequently, if one object modifies its heap data, the heap data of the other object will be unaffected. As importantly, if objB were to go out of scope before objA, there would be no negative impact on objA because objA would continue to point to memory that it solely owned.

As a constructor, the copy constructor has no return type (not even void) and bears the same name as the class. It takes one passed parameter (the source for copying) that is an object of the same type. All primitive, non-pointer fields can be immediately copied because separate memory has automatically been allocated for these fields. Memory has also been allocated for data members that are pointers but, as addresses, pointers should not be simply copied if data sharing is not warranted. Additional heap memory must be allocated before data values can be copied from the source object's heap memory to the newly constructed object's own heap memory.

The copy constructor is invoked to construct a new object; the assignment operator is invoked through an existing object. The compiler automatically generates an assignment operator that performs only a bitwise copy, resulting in unintentional aliasing if any data members are pointers. The assignment statement 'b = a' invokes the assignment operator through object b, so the address of b, '&b' is implicitly passed as the this pointer; a is explicitly passed as a parameter. In Example 3.2, b = a is equivalent to goodMM::operator=(&b,a). Obvious self-assignment, such as 'b = b', is usually optimized away by modern compilers. Since all cases of self-assignment are not evident due to aliases, the this pointer is used to check for self-assignment. Like the copy constructor, the assignment operator must allocate new heap memory but must also deallocate the heap memory owned by b.

Copying in C++ is onerous for the class designer but not for the client. The compiler automatically provides a default assignment operator and a default copy constructor which both perform bitwise copying. To avoid shallow copying, the class designer must define or suppress both the assignment operator and the copy constructor. Copy suppression simplifies class design but prevents the client from using call by value or the assignment operator. To correctly manage heap memory allocated internally in an object, the C++ class designer must also define the destructor. Example 3.2 displays a C++ class with properly managed internal memory: a destructor, copy constructor and overloaded assignment operator are all defined. goodMM objects allocate heap memory internally without memory leaks or hidden aliasing; call by value and assignment are both supported.

### Example 3.2 C++: Copy Constructor, Overloaded Assignment Operator

```
// good MemoryManagement: destructor and deep copying
class goodMM
{   private:
        int*  heapData;
        int   size;

        void copyData(const goodMM& source)
        {       size = source.size;
                heapData = new int[size];
                for (int j=0; j < size; j++)
                        heapData[j] = source.
                        heapData[j];
        }
    public:
        goodMM(unsigned s = 100)
        {   size = s;
            heapData = new int[size];   }
        // DEEP copying supported: copy constructor
        goodMM(const goodMM&  x)    { copyData(x); }

        // DEEP copying supported: overloaded
        // assignment operator

        // if not self-assignment:
        // delete old lhs memory
        void operator=(const goodMM& rhs)
        {       // address comparison efficient,
                // no self-assignment
                if (this == &rhs)      return;
                delete[] heapData;
                copyData(rhs);
        }

        // destructor deallocates heapData
        ~goodMM()                { delete[] heapData; }
};
```

The copy constructor and overloaded assignment operator are structured similarly: both copy primitive data directly, allocate heap memory of the same size as that held by the source object and then copy data values from the source heap memory. Hence, it is convenient to use a common private utility method, such as copyData in Example 3.2. The goodMM object which is the source for copying is passed by const reference to avoid the overhead of allocating and initializing a local copy. const provides security: any attempt to alter the formal parameter in the body of the function would cause a compilation error. The copy constructor is invoked for the construction of a new object. Without an existing object, there is no 'old' heap memory to deallocate. The assignment operator is invoked through an existing object, and, thus, to prevent a memory leak, must deallocate the 'old' heap memory associated with the object through which it was invoked.

If copying is not desired, as it is not for large data collections, a C++ class may suppress copying. Example 3.3 displays a C++ class with a destructor to prevent leaks but no support for copying. The copy constructor and overloaded assignment operator are declared private. Since private methods cannot be externally invoked, no implementation must be provided. The client can allocate objects but cannot copy from one object to another (via call-by-value or assignment). If client code attempts to assign one copySuppress object to another, the compiler generates an error, as shown in Example 3.3. copySuppress objects allocate heap memory internally without the possibility of memory leaks or data corruption because call-by-value and assignment are not supported. With C++11, copying may be suppressed by setting the copy constructor and overloaded assignment '=delete'.

If copying is not desired, why bother to declare the copy constructor or overloaded assignment operator, labelling them as private or '=delete'? If copying is not defined or suppressed, the compiler provides default versions which perform only bitwise copying. Such shallow copying produces unintentional aliasing and potential data corruption when the data members so copied are addresses. *Class design must explicitly suppress the copy constructor and the overloaded assignment operator to prevent the compiler's provision of default versions.*

### Example 3.3 C++: Suppressed Copying

```
// copying suppressed: private copy constructor
// and overloaded=
//    no need to define suppressed,
//    private methods in. cpp file
class copySuppress{
   private:
         int*    heapData;
         int     size;
```

```
        // copying suppressed
        copySuppress(const copySuppress&);

        // assignment suppressed
        void operator=(const copySuppress&);

  public:
        copySuppress(unsigned s = 100)
        {  size = s;    heapData = new int[size];  }

        // destructor deallocates heapData
        ~copySuppress() { delete[] heapData; }
        ...
  };

//client code - OK, invokes constructor
copySuppress c(20);
//client code: ERROR cannot invoke copy
//constructor
copySuppress d(c);              // compiler error   #1
//client code: ERROR, pass by value needs public
//copy constructor
void noLeakyFn(copySuppress x) // compiler error  #2
{              ...                          }
//client code: ERROR, assignment needs public
//operator=
d = c;                          // compiler error  #3
```

The sample client code in Example 3.3 declares copySuppress objects and then triggers compilation errors by attempting to copy copy-Suppress objects. In statement #1, the compiler attempts to allocate a new object d as a copy of existing object c but finds the copy constructor declared private in the copySuppress class and thus emits an error. With pass-by-value attempted in statement #2, the compiler again finds the copy constructor declared private and again emits an error. A compiler check of assignment in statement #3 finds a private operator= in the copySuppress class and, again, the compiler complains.

By default, copying is shallow, possibly leading to errors due to unintentional aliasing. *Shallow copying often may not be apparent.* When two objects point to the same heap memory and both objects assume ownership, data corruption is possible, in any language. Deep copies are more expensive but are safer because each objects points to and owns distinct heap memory. C++11 supports **move semantics**, examined in Section 4, for efficient and safe copying.

## 3.3  UNSEEN ALIASING

When C dominated software development, the need to manage memory was overt but often incompletely addressed. With the introduction of C++, the object-oriented paradigm came into vogue. *Objects encapsulated dynamic memory allocation and obscured the need for managing memory.* If class designers failed to manage memory correctly, memory leaks occurred even when clients 'followed the rules'.

Consider Figure 3.4, which displays the effects of pass by value: a is the formal parameter and b is the actual argument passed at the point of call. If the class does not define a copy constructor, the compiler provides the default bitwise copy constructor. With shallow copying, the formal parameter a accesses the heap memory allocated to the actual argument b, as seen in the first diagram, violating the security of pass by value. Moreover, when the function terminates, a goes out of scope. If a destructor is defined, as it should be, the destructor deallocates the heap memory

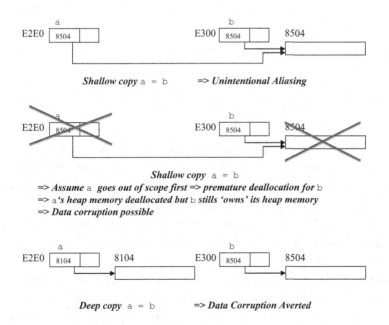

FIGURE 3.4   Shallow Copy versus Deep Copy.

that a 'points to'. But this memory is the heap memory 'owned' by b! But b does not know that its heap memory has been released and could be re-assigned to another 'owner'. Data corruption is now possible, if b, under the impression of continued ownership, alters 'its' heap memory. The desired deep (true) copy, as displayed by the third diagram in Figure 3.4, results when the class designer defines the copy constructor appropriately.

The function definitions in Example 3.4 look correct but are not. Memory allocation should be seamless (and handled by the compiler) because all variables in this example reside in function stack frames. whatIsWrong() looks innocuous, and it does nothing. The formal parameter, of type hiddenHeap, is passed by value and so the actual argument is copied to localVar. What could be incorrect? If the fault is not in the client code, then the hiddenHeap class definition is suspect. A class without any internally allocated heap memory will not leak memory but a class that internally allocates heap memory may.

**Example 3.4  Why Memory Leaks?**

```
// function code looks correct
void whatIsWrong (hiddenHeap   localVar)
{        int local = 42;      }

void howLeakMemory ()
{        hiddenHeap     steal;
         hiddenHeap     share;

         steal = share;
         return;
}
```

If hiddenHeap encapsulates a pointer that holds the address of heap allocated data then the class design should include a destructor and either suppress copying or support deep copying. Say hiddenHeap provides a destructor but fails to explicitly suppress copying or support deep copy-ing. Then the compiler generates a default, bitwise copy constructor which is invoked when a hiddenHeap parameter is passed by value into whatIsWrong(). Consequently, the formal parameter localVar addresses the same heap memory as the actual argument because shal-low copying establishes an alias by copying only address values. Data corruption is not a concern within function scope because the function does nothing; localVar does not alter the actual argument. True, but the invocation of the hiddenHeap destructor upon function exit will deallocate the heap memory that localVar references, which is the same memory that the actual argument references. This setup is the same as that illustrated by objects a and b in the first two diagrams of Figure 3.4. Data corruption is possible, an outcome dependent on use of the actual argument after the return from function whatIsWrong().

The second function, howLeakMemory() assigns one locally allocated (stack) object to another. Any memory leak then must be associated with the assignment operator. Object steal assumes the values from share. If steal contains a pointer to heap data, that address is overwritten unless an overloaded operator= has been defined to support deep copying. When the address of heap memory is lost before delete is called then the heap memory cannot be reclaimed. The memory leak in howLeakMemory() mimics that illustrated by objects objA and objB in Figure 3.2.

Clients cannot easily detect or correct data corruption or memory leaks that arise from a poorly designed class. Example 3.4 suggests that the hiddenHeap class does not have proper support for call by value or for assignment. In other words, the hiddenHeap class is missing a copy constructor and an overloaded assignment operator. Scott Meyers authored definitive guidelines on C++ class design, including thorough coverage of C++ class design for memory management. Please consult one of his texts.

### 3.3.1 C# Cloning to Avoid Aliasing

In C#, the default (bitwise) copying automatically establishes aliases since objects are references. For deep copying, C# class designs rely on cloning which requires casting. Historically, a recommended C# class design was to override Clone(), a method implicitly associated with every C# object (as is MemberwiseClone()). However, Clone() returns a (generic) object, requiring the caller to cast the return value back to the desired type. Such type reclamation may not be intuitive and shifts responsibility for type consistency to the client. A recent design preference is to internalize the casting process and provide two public class methods for copying – DeepCopy() and ShallowCopy(). Each method clones appropriately (a true copy of or a shared reference to an encapsulated subobject) and then casts the generic object back to the class type. Since both copy routines reclaim the appropriate type, the client does not bear any responsibility for casting. Example 3.5 outlines this approach with corresponding memory diagrams in Figure 3.5.

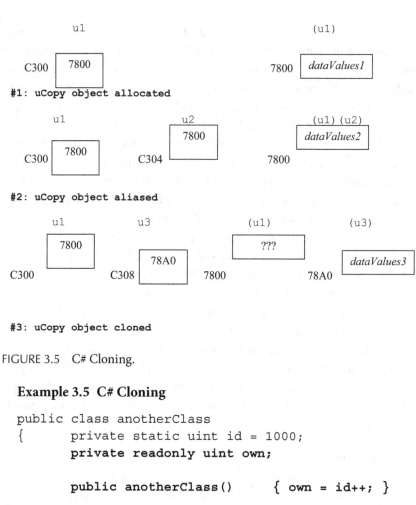

#3: uCopy object cloned

FIGURE 3.5    C# Cloning.

## Example 3.5  C# Cloning

```
public class anotherClass
{       private static uint id = 1000;
        private readonly uint own;

        public anotherClass()      { own = id++; }

        public uint Own { get => own; }
}

public class uCopy
{       private        anotherClass  address;

        public uCopy()
        { address = new anotherClass(); }
        // deep copy: heap memory allocated for
        // true copy
        public uCopy DeepCopy()      // internal cast
        {       uCopy local = (uCopy) this.
                MemberwiseClone();
                local.address = new anotherClass();
```

```
            return local;
        }

        // shallow copy: distinct objects have same
        // address
        public uCopy ShallowCopy()
        {   return (uCopy) this.MemberwiseClone(); }
}

// client code WITHOUT casting
// #1: uCopy object allocated
uCopy   u1 = new uCopy();
                            // embedded 'own' id of 1000
// #2: shallow copy, aliased with u1
uCopy   u2 = u1;
                            // embedded 'own' id of 1000
// #3: shallow copy, aliased with u1
uCopy   u3 = u1.DeepCopy();
                            // embedded 'own' id of 1000
// #3: deep copy; DISTINCT object
uCopy   u4 = u1.DeepCopy();
                            // embedded 'own' id of 1001
```

Data corruption arises from inappropriate aliasing, in any programming language, and its effects may be immediate or delayed. If two objects hold the address of the same heap memory, either object can alter the shared data, thus 'corrupting' that data for the other object. Class designers must determine whether or not copying should be supported, and, if so, whether deep or shallow copying is appropriate.

## 3.4 MOVE SEMANTICS

Appropriate design of suppressed, shallow, or deep copying assures correctness. Deep copying prevents aliasing but is more expensive than shallow copying, especially for large objects. Can one avoid this expense? Yes, by transferring ownership. C++11's move semantics offer the ability to do so. Move semantics are an implicit optimization technique that reduces the number of allocator (new) and deallocator (delete) calls by assuming the memory of an expiring temporary.

Consider return by value for *a C++ object that contains a pointer to heap memory.* The copy constructor is invoked to copy the local (stack) object into the temporary returned to the caller. When scope is exited, destructors are invoked for all objects resident in the function's stack frame, releasing all heap memory owned by stack objects. A shallow copy of the local (expiring) object to the return object would result in the caller's return object indirectly addressing released heap memory, leading potentially to data corruption. Thus, shallow copying is not merely inadequate but dangerous. A deep copy of the local (expiring) object to the return object though is unwarranted. Deep copying invokes the new operator to allocate additional heap memory for a true replica so that the target copy (return object) does not interfere with the source for copying (the expiring local object). But when the local object goes out of scope, its destructor is invoked and thus its heap memory is deallocated. Hence, it is not possible for the object returned upon function exit to interfere with the expiring local object.

Transferring ownership from a local object to the return copy is a viable alternative when the source for copying is expiring. Move semantics embodies the transfer of heap memory ownership. The move constructor assumes the data content of the source object, avoiding a call to the new operator and the copying of data values. With ownership transfer, the expiring object has released its memory so there is no need to call the delete operator. The move constructor saves one allocator/deallocator pairing, making copying more efficient while preserving data integrity.

Similar inefficiencies arise when using deep copying for assignment. Consider the assignment statement "c = a + b" – which is actually two operations: construction of the composite of a and b; and then, copying this composite to existing object c. A temporary is returned from the operator+ method and is immediately assigned to c. Again, a shallow copy is inappropriate when objects internally reference heap memory: 1) copying only the values of data members that are addresses will overwrite c's handle to heap memory yielding a memory leak; 2) destructor invocation when the temporary goes out of scope will yield a dangling pointer for c, possibly leading to data corruption. A deep copy is inefficient because, again, the source for copying is an expiring temporary so there is no need for it to retain heap memory. The move assignment operator swaps memory with the temporary, preventing data corruption, and achieving efficiency. No call to new is needed for c when it assumes the temporary's memory. No call to delete is needed for c because the temporary now references c's pre-assignment heap memory so deallocation is guaranteed when the temporary goes out of scope. Again, the overhead of one allocator/deallocator pairing is avoided.

As shown in Example 3.6, the move constructor and move assignment operator are distinguished from the standard copy constructor and assignment operator by '&&'. Also note that the passed parameter (source object) cannot be 'const' because it must release ownership of its memory.

**Example 3.6 C++11: Move Semantics**

```cpp
// copying avoided: assume data of source
// reference
//     define move constructor and move assignment
//     operator
class copyAcquire{
   private:
        int*    heapData;
        int     size;

public:
        copyAcquire(unsigned s = 100)
        {   size = s;   heapData = new int[size];   }

        // (deep) copy constructor
        copyAcquire(const copyAcquire&  a);

        // copying avoided via move constructor&&
        //          source object c yields ownership
        copyAcquire(copyAcquire&&  c)
        {   size = c.size;      heapData = c.heapData;
            c.size = 0;         c.heapData = nullptr;
        }

        // assignment with deep copy
        copyAcquire& operator=(const copyAcquire& a);

        // move assignment&& exchanges ownership
        copyAcquire& operator=(copyAcquire&&  c)
        {          swap(size, c.size);
                   swap(heapData, c.heapData);

                   return *this;
        }

        // destructor deallocates heapData
        ~copyAcquire()
        { if (heapData)          delete[] heapData; }
};
```

How do move semantics work? The move constructor assumes ownership from the source object. The newly constructed object acquires the heap address held by the source object (avoiding a call to new) and the source object's heap pointer is set to null. The move assignment operator swaps the heap memory of the left-hand side of an assignment statement (**lvalue**) with that of the right-hand side of an assignment statement. When the source object subsequently goes out scope, its destructor deallocates the heap memory originally owned by the lvalue. Ownership transfer reduces run-time overhead by reducing allocator and deallocator calls. The parameter passed to a move constructor or a move assignment operator is an expiring temporary with no need to retain its memory.

The compiler is responsible for invoking the move constructor in place of the copy constructor or the move assignment operator instead of the assignment operator. The compiler identifies expiring temporaries and distinguishes between references that permit assignment (lvalues) and those that do not. The compiler determines the legality of transferring memory ownership and checks for class support of move semantics.

A move constructor and a move assignment operator should be defined whenever the copy constructor and an overloaded assignment operator are defined for deep copying. The compiler resolves which constructor or assignment operator to call, based on whether memory may be assumed. Hence, move semantics are safe and efficient.

## 3.5 HANDLE: C++ SMART POINTERS

A **handle** provides the means to access data. A variable is an abstraction of an assigned memory location but provides direct access to data through its name. Pointers and references hold data that is interpreted as addresses and thus provide an indirect means to access data (though indirection is not obvious when using a reference). A single piece of data may have multiple handles, as is evident with pass by reference.

A classic example of a handle is the **smart pointer**, first embodied by the C++ auto _ ptr class (now deprecated) in the STL (Standard Template Library). A smart pointer is a wrapped pointer used to control access to and manage memory. By wrapping a raw pointer inside a class, with a defined destructor, memory leaks are averted. The term 'wrapper' has colloquially subsumed the 'term' handle. Formally though, a smart pointer is a handle rather than a wrapper; a smart pointer does not wrap up an interface. By overloading operator->() and operator*(), calls to the functionality of the wrapped type may be forwarded transparently (see

Chapter 8). Thus, the client may use the wrapped pointer as if it were a pointer typed to encapsulated type.

C++11 deprecated auto _ ptr, replacing it with three generic smart pointer types that encapsulate a raw pointer: unique _ ptr assumes sole ownership; shared _ ptr models shared ownership via reference counting; weak _ ptr provides (shared) access but cannot prevent deallocation. This refinement distinguishes different ways to control access to memory resources: a simple wrapper, a reference counter, a secondary reference. The notion of an encapsulated pointer, as a means to safeguard memory management remains. A unique _ ptr triggers invocation of a destructor when it goes out of scope, and in so doing, prevents memory leaks. A shared _ ptr triggers invocation of a destructor only if the reference count drops to zero. A weak _ ptr cannot trigger invocation of a destructor.

C++11 smart pointers:

1. unique _ ptr **automates transfer of ownership**

    a. Only one owner => exclusive access to wrapped raw pointer

    b. Deallocation automatic when unique _ ptr goes out of scope

2. shared _ ptr **automates reference counting to track aliases**

    a. Shared access to wrapped raw pointer

    b. Deallocation only after last reference released

3. weak _ ptr **facilitates transient access**

    a. shared access to wrapped raw pointer (may share with shared _ ptr)

    b. No ownership claims => no effect on retention

Example 3.7 illustrates a wrapped pointer, grabMemoryPtr, whose constructor assumes ownership of the memory referenced by the pointer parameter. The passed pointer is zeroed out, preventing the caller from using the raw pointer after it has been wrapped. When a grabMemoryPtr object goes out of scope, its destructor is automatically invoked so the deallocation of heap memory is guaranteed. Overloaded operators provide transparency so the client may use the wrapped pointer as if it were a raw pointer. This design outlines the conceptual basis for unique _ ptr. Smart pointers though are generic and include restrictions on copying.

**Example 3.7 Wrapped Pointers in C++**

```
// Destructor invocation prevents memory leaks
// Constructor assumes ownership of memory
// addressed by parameter
//    =>   client cannot use raw pointer thereafter
class grabMemoryPtr
{       SomeType*       ptr;
    public:
        // pointer passed by reference:
        // value zeroed out
        grabMemoryPtr(SomeType*& p) : ptr(p)
        { p = 0;}

        ~grabMemoryPtr()          { delete    ptr; }

        // forward calls transparently
        SomeType*  operator->() { return ptr; }
        SomeType   operator*()    { return *ptr;
}
};
```

## 3.5.1 unique_ptr

Conceptually, unique _ ptr wraps a raw pointer that holds the address of a heap object and claims sole ownership of that 'unique object'. With minimal overhead, unique _ ptr<T> manages a pointer to object of type T. When a unique _ ptr goes out of scope, its destructor is invoked, guaranteeing deallocation of the addressed (heap) object. When an exception is thrown after a call to new but before its matching delete, raw pointers may leak memory. unique _ ptrs do not leak memory upon premature exit due to a thrown exception. Since unique _ ptrs cannot be copied, ownership is constrained. Unintended sharing or aliasing is prevented: there are no copy constructors or assignment operators for unique _ ptrs. To pass unique _ ptrs by value, unique _ ptrs must first be 'move'd. unique _ ptrs may easily be passed by reference.

Example 3.8 illustrates the effective use of unique _ ptr to prevent memory leaks. Novice programmers may not anticipate the leakage implied by statement #1 in the twoLeaks function. The allocator is called via the new operator and the address of a heap object is returned. However, this address is not stored in a pointer: it is dereferenced and the values of the heap object are copied into a stack object. The address to the heap object is lost. Hence, there is a memory leak because there is no pointer through which to call delete. The second leak in twoLeaks occurs because of a premature exit: the function is exited prior to the call to delete.

**Example 3.8  unique_ptr Assumes Ownership => Leak Prevented**

```
#include <memory>
void  twoLeaks(int  x)
{      // #1: value of heap memory copied to stack
       // object
       //        address of heap object lost => memory
       //        leak
       SomeType stackObj = *(new SomeType(x));

       SomeType*      trackAddress = new SomeType(x);

       // #2:exit before delete, by DESIGN or THROWN
       // EXCEPTION
       int  preMatureExit;
       cin >> preMatureExit;  // IO error => exception
       if (preMatureExit)   return;  // coding error

       delete trackAddress;
}
void  noLeakAnonymous()
{      // #1 exit scope => memory addressed by
       // p & q deleted
       unique_ptr<SomeType> p( new SomeType(x));
       unique_ptr<SomeType> q( new SomeType(x));

       // #2: premature exit ok: no delete needed
       int  preMatureExit;
       cin >> preMatureExit;
       if (preMatureExit)   return;
}
```

Move semantics ensure consistency by avoiding copying. Consider transferring an object into an STL container, v.push _ back(std::move(p)), via unique _ ptr  p. The move constructor releases ownership of p's wrapped pointer to the container v and sets the internal pointer owned by p to null. Subsequent use of p may trigger compilation errors or generate a null pointer exception at run-time. In contrast, returning unique _ ptr from a function needs no special code.

### 3.5.2 shared_ptr

Shared access broadens the scope of and dilutes ownership. Persistence may be controlled by associating a counter with an object which is set to '1' upon initial allocation. Each subsequent alias ('copy') increments the reference count; each handle thus tied to the data item decrements the reference count when it goes out of scope; the data is then deallocated when reference count is '0'.

Conceptually, shared _ ptr reflects shared access, and thus suppression of copying. A static count tracks how many shared _ ptrs point to the same heap object. When the last shared _ ptr goes out of scope, this reference count drops to zero and triggers the destructor so no memory leaks. shared _ ptrs simplify memory management. It is safer to return a shared _ ptr than a raw pointer from functions: the caller is not obligated to deallocate any returned data (and need not worry whether delete or free be called). shared _ ptrs make memory management more efficient since copying data is avoided. The 'creation' and 'release' of objects is faster because only the first allocation and the last deallocation involve a constructor or destructor; there is only one allocation/deallocation for the shared object and for its reference counter. STL containers accept shared _ ptrs.

For shared access, function make _ share hides the new operator and: 1) allocates contiguous memory for an object and its reference count; 2) invokes the instantiated class constructor, forwarding any arguments; 3) returns a shared _ ptr to newly created object. make _ unique and make _ shared are preferred over new and malloc; both encapsulate deallocation and eliminate the need for delete and free.

Carefully distinguish between the need for unique _ ptr versus shared _ ptr. Is sharing necessary? Should all indirect access be equal? How is data integrity ensured? Exclusive access mandates unique _ ptr. A common recommendation is to use a unique _ ptr, knowing that one can convert to a shared _ ptr, if necessary. In particular, caution is advised for concurrent software. shared _ ptr permits sharing of a resource thorough multiple pointers but does NOT enforce thread safety; wrapping an object in a shared _ ptr does not make it thread safe. Access to the shared resource managed by shared _ ptr must be controlled. To prevent data corruption, synchronization must be addressed or a unique _ ptr should be used.

### 3.5.3 weak_ptr

weak _ ptr, is a weak reference to an object managed by shared _ ptr. Like Java/C# weak references, weak _ ptr prevents cyclic references that impede memory reclamation. As a reference with no ownership claim to a shared heap object, a weak _ ptr does not contribute to the reference count of managed object. The referenced object is deallocated only after all shared _ ptrs release ownership, regardless of how many weak _ ptrs still reference the released resource. Since a weak_ptr has no control over deallocation, always check if a weak _ ptr is valid before use: a.expired()returns true if weak _ ptr a points to released resource.

Any weak _ ptr is easily converted to a shared _ ptr, which provides temporary ownership and safe access to an object. Consider using weak _ ptr when ownership of a resource is not needed, when control of object lifetime should not be a responsibility, or when cyclic references may be problematic. To confine responsibility, and to prevent memory leaks and data corruption, replace raw pointers with unique _ ptr or shared _ ptr.

### 3.5.4 Usage

Smart pointers were designed for safety, ensuring that heap allocated memory does not leak. As a wrapper, each smart pointer has a destructor which will automatically execute when scope is exited even when an exception is thrown. Smart pointers provide automatic resource management, cleanly compensating for failure to deallocate heap objects (via delete or free) as well as unreachable deallocation requests due premature exit. Available via <memory>, smart pointers provide transparent access to a wrapped type so that clients may manipulate smart pointers in the same manner as raw pointers.

1. Guidelines:Wrap raw pointers (use smart pointers) to ensure safe and efficient use of memory

2. Replace auto _ ptr in legacy code with unique _ ptr

3. Use unique _ ptr.reset() for deallocation => no need to unwrap the pointer

4. Avoid unique _ ptr.release() which transfers ownership (and deallocation responsibility) to the caller

5. Avoid shared _ ptr.get() which extracts the handle to a shared raw pointer

=> caller should not delete raw pointer (but cannot be prevented from doing so)

Our brief coverage of smart pointers only highlight new language features to support correct memory management. The distinction between sole ownership and shared references should have been reinforced. For implementation expertise, consult http://en.cppreference.com/w/cpp/memory for more detail

## 3.6 OO DESIGN PRINCIPLE

A class design with well-defined copy semantics minimizes client responsibilities for data integrity. The **Principle of Least Knowledge** (also known as the Law of Demeter) exalts information hiding and directs the client only to the public interface.

**Principle of Least Knowledge (PLK).** Every object should assume the minimum possible about the structure and properties of other objects.

PLK implies that *a client should know as little as possible about objects* and is supported by Programming By Contract. The client should be guided by the class and interface invariants, not the implementation invariant. Pre and postconditions direct the client to verify passed data and to track only the publicly relevant notion(s) of object state. Client contractual responsibilities should center on externalities not implementation details. As long as the public interface remains stable, internal modifications to a type definition do not affect the client.

## 3.7 SUMMARY

This chapter examines correctly managed memory within a class design when copying is supported. Class design should explicitly consider copying techniques, for both correctness and efficiency. Internal resource management reduces client responsibility. Memory is abstracted but remains a design responsibility, especially when data integrity, performance, and scalability are critical requirements. Data corruption arises from poorly tracked or safeguarded ownership of allocated data. Contractual design thus should notify a C++ client of suppressed copying and a C# client of cloning options. Options are noted below.

Design Choices for Copying:

1. Shallow Copying –        *bit strings copied from source to destination*
                            **default process sufficient for primitives**

2. Deep Copying (or Cloning) – *avoid bitwise copy for addresses values*
                            **replica requires own heap memory**

3. Suppressed Copying in C++ – *no replication*

   a. private (or 'delete'd) copy constructor and operator=

   b. controlled wrapper – smart pointer

   c. move semantics – compiler transfers ownership only when safe

Class design, with internal copy semantics and published contractual expectations, supports abstraction and maintainable client code. Nonetheless, the overhead of memory access might still be significant. Since data and instructions are transferred in blocks from memory, overhead is reduced when accessing contiguous data or sequential instructions. Software design principles for efficient data manipulation include: minimize I/O, use constants where appropriate, reduce the generation of temporaries, avoid unneeded copying, and transfer ownership. Table 3.1 summarizes general design techniques for efficiency. This chapter also examined compiler optimizations for efficiency as supported by C++ move semantics.

TABLE 3.1    Design Techniques for Efficiency

|  | Cost | Response |
|---|---|---|
| **I/O** | Memory Access | Minimize |
| **Function Call** | Call & Return | Inline (compiler optimization) |
|  | Loss of Spatial Locality | Use virtual methods carefully |
| **Copying** | Time | Aliases, Call by Reference |
|  | Space | Smart Pointers |
| **Temporaries** | Allocation/Deallocation | Avoid Generation |
|  | Fragmented Heap | Call by Reference |
|  |  | Move Semantics |

## 3.8 DESIGN EXERCISES

This chapter's first exercise is to redesign the C++ `feeLedger` class from Chapter 2, using the same internal array structure, to support deep copying, correctly and efficiently. The second exercise is to redesign `feeLedger` again, using STL `vector`, to 'automatically' support deep copying, correctly, and efficiently. Appendix B.3 presents and analyzes sample solutions.

**DESIGN INSIGHTS**

*Software*

Ownership of persistent data may be transferred

Copy semantics

Deep copy versus shallow copy

Suppression as a viable option

Affect data usage (efficiency and correctness)

Require explicit class design decisions

Data corruption errors hard to track

*Software Design*

C# objects are references

Data persistence automatically obtained

Copying shallow by default

Client relies on class design and public methods for deep copying

C++ objects encapsulate dynamic memory allocation

obscure need for direct memory management

may suppress copying by design

may support compiler optimizations

## CONCEPTUAL QUESTIONS

1. When is heap memory used?

2. Why is tracking the ownership of (heap) memory difficult?

3. What is the primary difference between deep and shallow copying?

4. When is it appropriate to suppress copying?

5. What are the differences between C++ and C# copying?

6. Identify C++ class design requirements for safe and efficient use of memory.

# II

## Strategic Type Coupling

## CHAPTER 4

# Composition

## CHAPTER OBJECTIVES

- Define OOD relationships

- Illustrate composition and containment

- Examine association, ownership and cardinality

- Introduce Dependency Injection

## 4.1 OBJECT-ORIENTED RELATIONSHIPS

A famous quote from Aristotle, "the whole is greater than the sum of its parts", emphasizes the power of combination. By reusing and combining types, class designers may expediently construct new types and new interfaces. A central design question is how to do so. An implied design responsibility is to manage the dependency of the whole on its parts.

OOD defines different relationships (composition, containment, and inheritance) that determine the form and flexibility of reuse. How types are connected – association, cardinality, and ownership – differentiate design options. An association between two objects may be temporary, stable, or for the lifetime of the primary object. Cardinality may reflect a one-to-one or a one-to-many relationship, may be defined at the class or object level, and may vary or be stable. Ownership implies that the primary object is responsible for a secondary object, requiring explicit decisions for allocation, release, replacement, or transfer of ownership.

Basic structural relationships are has-a (composition), holds-a (containment), and is-a (inheritance). Historic OOD discussion defined **aggregation** as a structure where the aggregate object contains many subobjects of the same type. *Aggregation addresses only form and not intent or effect.* For example, both a container and a building toy (such as a Lego set) may be described as aggregates. However, a container retains no dependency on the subobject type while a building toy is strongly dependent on its components. A container illuminates a holds-a relationship where there is little restriction on the type of subobjects held while a composite illustrates a has-a relationship with significant dependency on the subobject type. While structurally similar, holds-a and has-a may be distinguished via design details such as association, ownership, lifetime, and reuse of functionality.

The simplest relationship is none: two types do not interact. Next in simplicity is the uses-a relationship where one type uses another in a transient fashion such as call by value. Other relationships represent associations that are more enduring and suggest some type dependency.

## 4.2 CONTAINMENT (HOLDS-A)

Standard containers model the holds-a relationship well because there is no type dependency on the subobjects. A stack provides the same utility no matter what type of data held. A stack is well-defined when empty, full, or in-between. The operations of push(), pop(), clear(), etc. function in the same manner regardless of the type of data processed. The type of data stored provides no functionality and has little or no effect on containers. The holds-a relationship reflects little or no type dependency.

Example 4.1 portrays weak type dependency: a customer holds-a gift card. The no-argument constructor zeroes out the pointer defined to hold the address of a (heap-allocated) gift card, suggesting that a customer may operate without a gift card and that not all customers have a gift card. If a customer is well-defined without a gift card, then a customer may have zero gift cards and still function as a customer. A customer is not dependent on a gift card if gift cards do not drive core functionality, or if another item, such as a free shipping certificate, may replace a gift card.

Holds-a does not require ownership. The customer may not be responsible for the destruction of a gift card, especially if ownership is temporary. Disposal of a gift card may differ by design. If the customer is the sole owner of a dynamically allocated gift card, then the gift card should be 'destroyed' (reference zeroed out or destructor invoked) unless ownership is transferred out. Since the presence of a gift card is optional in the customer class of Example 4.1, any method that accesses the gift card must first test for existence, as is done in replace().

**Example 4.1  C++ Customer Holds-A Gift Card**

```
// transient ownership of subobject(s)
// implies memory management
//      => must provide destructor
//      => support or suppress copying
class Customer               // replaceable gift card
// handle only, no object yet
{       GiftCard*   c = 0;
   public:
           // assumption constructor: ownership
           // of transfer assumed
           Customer(GiftCard*& transfer)
           {      c = transfer;
                  transfer = 0;
           }

           // no argument constructor:
           // no gift card allocated
           Customer()   {   c = 0;      }

           // again ownership transferred in
           void replace(GiftCard*&    backup)
           // dispose existing card
           {       if (c) delete c;
                   c = backup;
                   backup = 0;
           }
           ~Customer()   {    if (c) delete c;      }
};
```

A container may hold objects, copies of objects or references to objects. The objects contained may be passed in and out, transferred, or destroyed (redeemed), yielding a fluctuating cardinality across the lifetime of the container. Logically, a customer may hold a positive number of gift cards, or none. If different gift card types are available (bonus, restricted by item or calendar date, etc.), the mix of gift card types held may vary over the lifetime of a customer. Only a temporary association exists between the customer object and the gift card object.

Independent of implementation language, *a containment relationship is flexible because cardinality, ownership, and association may vary.* Designs differ though because of implementation language. In C++, memory management must be addressed for any object with internally allocated heap memory. The class must track ownership so that all heap-allocated memory is deallocated before objects owning the heap-allocated memory go out of scope. In all languages, aliases should be tracked so that dead objects may be reclaimed and data is not corrupted.

Copying is an essential design decision. Often, it is undesirable to copy large collections either for data integrity or performance concerns. What are the effects of supporting or suppressing copying? What does a container hold: original data, duplicates, references? Copying may be more complex when data is referenced indirectly, that is, via a reference or a pointer. What is copied? – the address holder (reference or pointer), or the actual data values?

*Copy semantics should be an explicit design decision.* If a C++ class neither defines nor suppresses copying, the compiler generates a default copy constructor and overloaded assignment operator that yield shallow copies, and, thus aliasing and possibly data corruption. If no decision is made in C#, copying is also shallow. Recall the difference between shallow and deep copying as examined in Chapter 3.

## 4.3 COMPOSITION (HAS-A)

An intuitive example of composition is a signal that relies on sensors. A signal is activated when some number or proportion of its sensors are triggered. An alarm clock uses a timer as a sensor; a security light uses a motion detector, etc. Type dependency is clear: a signal is not well-defined without sensors – it is inoperable. In a has-a relation, the subObject provides key functionality and affects the state of the composing object; for example, sensors affect the state of the signal. A sensor in a failed state affects the functionality of the signal. If all sensors are off, then the signal is effectively off, etc.

In composition, the association between the composing object and its subobject(s) is usually stable. With fixed cardinality, the number of subobjects may be defined by design or set upon object construction but it does not vary within an object's lifetime. For example, if a signal is designed with eight sensors, then all signal objects have eight sensors. In contrast, as Example 4.2 demonstrates, the number of subobjects may vary from object to object when cardinality is defined in the constructor but not altered thereafter. Signal objects instantiated from the same class definition may have different numbers of encapsulated sensors but each signal object has a fixed number of sensors for its lifetime. Conveniently, C#'s keyword readonly prevents change after initialization. The implementation invariant should record the stability of subobject cardinality, especially when no keyword prevents change.

**Example 4.2 C# Stable Number of subObjects**

```
// cardinality fixed in constructor
public class Signal
{       private Sensor[]            trigger;
        private readonly  uint      numSensors;

        public Signal(uint num)
        // bound number of sensors
        {       numSensors = num % 10;
                trigger = new Sensor[numSensors];
                for (int k = 0; k < numSensors; k++)
                        trigger[k] = new Sensor();
        }
        ...
}
```

A subobject in a has-a relationship is typically considered owned by the object and not shareable – usually but not always. An object may depend on a subobject that is a shared resource. In which case, management of the shared resource is likely external to the object (as with a shared_ptr). Control of a shared resource (via reference counting, locks, etc.) is then not the responsibility of the composing object. Although ownership of subobject(s) typically resides with the object, any composition design may transfer ownership out to a caller or may assume ownership from a caller.

Composition reinforces internal control. The state of a subobject may affect its utility. Is a file read-only? Is a password valid? Composition buffers the client from verification details. Composition affords much flexibility. Subobject volatility may be handled by replacement or excess capacity. Not all subobjects in a has-a relationship must be used. A signal with multiple sensors may sample or rely on only a subset of sensors. One or more sensors may be off or may fail (or may not be instantiated) but the signal can still function if enough sensors remain operational.

### 4.3.1 Modification

Composition preserves control over the replacement and state change of an encapsulated data member. Example 4.3 illustrates a device that uses a (charged) battery. When the device is on, all functionality is supported but the battery charge is drained. Turning off a device saves its battery but disables functionality. Dead batteries may be recharged or replaced. State changes, such as recharging, often require less overhead than replacement but may not always be feasible.

**Example 4.3  C# Modification of Encapsulated Data Member**

```csharp
public class Battery
{      private bool on = true;
       private double charge;

       public Battery(double chrg = 1212.12)
       { charge = chrg; }

       public bool On { set; get; }
       public double Charge
       {      get => charge;
              set;
       }
}

public class Device
{      private Battery              duraCell;

       public Device(uint          charge)
       {       duraCell = new Battery(charge);           }

       private bool                flipOffOn(bool on)
       {               // #1: no state changed needed
              if (on == duraCell.On) return on;

              // #2 request to turn off always met
              if (!on)
              {                    duraCell.On = false;
```

```
                    return false;
        }
        if (duraCell.Charge < 100)
                duraCell.Charge *=10;
        duraCell.On = true;
        return true;
        }
    }

    public void                 operate()
    {           if (!duraCell.On)                return;
            flipOffOn(true);
            // ...perform requested operation
            // using battery
    }
}
```

Class control of subobject state and existence extends to realistic examples where resources are files, passwords, connections, etc. Composition reuses rather redefines the utility provided via the subobject's public interface. The composing class must compensate for any deficiencies in the subobject's public interface, including restoration to a usable state. Thus, replacement is often employed. For example, if a battery is not rechargeable then a device may need to replace its battery data member.

### 4.3.2 Replacement

Example 4.4 provides a C++ implementation of Example 4.2 that retains the design of establishing a stable number of sensors in the constructor (while suppressing copying). Additionally, the wholesale replacement of sensors is provided. Replacement may be designed to preserve or alter cardinality. Recall that the allocation of heap memory internal to a C++ class requires an explicit design decision for copying – suppress or support with deep copying and move semantics.

Replacement strategies preserve utility when subobjects can no longer provide needed functionality, as when sensors malfunction or batteries die. Replacement may change the number of encapsulated subobjects. For example, more sensors may be added in order to increase surveillance (e.g. motion detectors); fewer may be used to decrease power consumption. Replacement should be designed carefully and controlled:

requests for replacement may be rejected, the number of replacements may be limited, etc.

### Example 4.4 C++ Variable Number of subObjects with Replacement

```
class Signal
{        Sensor*              trigger;
         unsigned             numSensors;

         // copying suppressed
         Signal(const Signal&);
         void operator=(const Signal&);

         void validateState()
         {        int     defects = 0;
                  for (int k = 0; k < numSensors; k++)
                          if (trigger[k]->dead())
                               defects++;
                  if (defects > numSensors/2)
                          replace(numSensors);
         }
    public:
         Signal(unsigned int num)
         {        numSensors = num % 10;
                  trigger = new Sensor[numSensors];
         }
         ...
         ~Signal()         {        delete[] trigger;      }

         void replace(unsigned num)
         {        delete[]  trigger;
                  if (num != numSensors)
                       numSensors = num % 10;
                  trigger = new Sensor[numSensors];
         }
};
```

Resource acquisition should be securely managed. Internal control may limit replacement and may avoid (or postpone) the allocation of expensive resources, to improve efficiency. Example 4.5 illustrates internal management of a passcode. Theft or expiration may invalidate a passcode so replacement is anticipated. Class design may restrict conditions for replacement and may limit the number of replacements.

### Example 4.5 C# Internal Replacement

```
public class passCode
{       public passCode(uint x)    { ... }
        public bool isValid()      { ... }
        public bool expired()      { ... }
}


public class Verifier
{       private     passCode        key;
        private     uint            id;
        private     uint            numReplace;
        private     uint            maxReplace;
        ...
        public Verifier(uint  id)
        {     key = new passCode(id);
              maxReplace = 1 + id % 1000;
        }
        ...
        public bool passOK()
        {     if (key.inValid())        return false;
              if (key.expired())
              {     if (numReplace >= maxReplace)
                        return false;
                    numReplace++;
                    key = new passCode(id);
              }
              return true;
        }
}
}
```

View replacement as a two-step process: release old resource; acquire new resource. Heap memory, locks, database connections, etc., are examples of resources that may be encapsulated and for which the replacement may be null. A key design decision is whether or not the client should be aware of subobject replacement. In Example 4.5, the use and replacement of passCode is completely internalized; the client is unaware of the subobject existence. Our coverage of Dependency

Injection later in this chapter will consider designs that externalize replacement responsibility.

### 4.3.3 Postponed Instantiation

An early employee of DEC (Digital Equipment Corporation) once quipped "the most reliable components are the ones you leave out". Composition affords the control of internal construction. A subobject may be instantiated upon object construction. Alternatively, construction of a subobject may be postponed until use (like just-in-time manufacturing – only instantiate (manufacture) when needed). When an object becomes responsible for subobject allocation (and possibly deallocation), copy semantics must be defined. Review Example 4.6, keeping in mind that internally allocated heap memory requires careful design in C++ (as shown in Chapters 2 and 3). Note that the handle to the dataset is zeroed out in the constructor, indicating that no dataset has been instantiated. Hence, class methods must test for existence before using the dataset.

**Example 4.6  C++ Postponed Instantiation of SubObject**

```
class justInTime
{       // need memory management:
        // destructor & copy semantics
        bigData*        dataSet;
        int             fakeStart;
        bool            alive;
        ...
    public:
        justInTime(int min = -1)
        {       dataSet = 0;
                fakeStart = min;
                alive = min > 0;
        }
        ...
        bool wideRange()
        {       if (!dataSet) dataSet = new bigData;
                return (dataSet->getMax() >
                        100*dataSet->getMin());
        }

        bool validRange()
        {       if (!alive)             return false;
```

```
            if (!dataSet)
                dataSet = new bigData;
            return (fakeStart < dataSet->getMin());
    }


    ~justInTime()
    {       if (dataSet)              delete dataSet; }
        ...
};
```

In client code, *many temporary objects are generated but not fully used.* Local objects, conditionally used in confined scope, and C++ temporaries, constructed via call by value, may not ever be used to invoke a method that relies on a subobject. In which case, there is no need to allocate subobjects. Postponed instantiation avoids allocating an unused resource, enhancing performance in any language. Postponed instantiation does not necessarily indicate a weak dependency. Existence checks must be added to all methods that depend on a subobject whose instantiation occurs after object construction.

In Example 4.6, `justInTime` objects that do not invoke `wideRange()` or `validRange()` will not incur the cost of allocating the encapsulated `bigData` data member. Any `justInTime` objects instantiated without a value or with a negative integer will not trigger the allocation of the encapsulated `bigData` data member either in the constructor or in `validRange()`. When a `justInTime` object goes out of scope without a `bigData` subobject allocated, the destructor need not call the deallocator.

Constraining heap allocation reduces run-time allocation costs for both C++ and C#. Additionally, deallocation costs are reduced in C++ and heap fragmentation in C#. Deferring object instantiation is a popular design choice because it often decreases overhead, especially when objects are instantiated but not fully used as is often the case for temporaries. Postponed instantiation can be viewed as on demand instantiation.

### 4.3.4 Echoing an Interface

Since a subobject is a private data member, its existence and interface are hidden from the client. This encapsulation decouples the client from the subobject, buffering the client from change, and from responsibility for the subobject. Software maintainability is thereby promoted. The client has little or no knowledge of encapsulated data members and cannot directly invoke any functionality published in the subobject's public interface. The class designer though can support indirect invocation of such encapsulated functionality by defining public methods to 'echo' subobject methods.

**Example 4.7  C++ Echoed Interface**

```cpp
class Echo
{       workerB          subObject;
        unsigned         countQueries;
        bool             passThru;
    public:
        // initializer list: invoke
        // non-default constructor
        Echo(int seed = 10): subObject(seed)
        {       countQueries = 0;
                passThru = seed > 0;
        }
        ...
        double work()       // #1 immediate pass thru
        {       return 2*subObject.work();   }

        // #2 preprocessing before pass thru
        bool    isValid()
        {       countQueries++;
                return subObject.isValid();
        }

        // #3 conditional pass thru
        int     sizeHive()
        {       if (passThru)
                    return subObject.sizeHive();
                return -1;
        }
};
```

The composition relationship of class Echo, as shown in Example 4.7, echoes a portion of the workerB interface. A client may invoke the methods work, isValid, and sizeHive through an Echo object, receiving the results obtained from the encapsulated workerB data member. The Echo class designer chooses which methods to echo and may do so conditionally, with or without pre and post processing. For example, size-Hive() will not return valid information if the flag controlling echoing is false. The class designer may refine behavior by adding code in the shell method, such as the example of doubling the 'work' value. Echoing an interface reuses an existing type while maintaining control over the invocation of methods so acquired.

Composition promotes internal control. Any composition design may choose to: modify subobject state, replace subobjects, defer instantiation, and/or echo a subobject interface. These design variants can be used singly or in conjunction with each other. With a stable interface, modification of internal design details has minimal impact on the client. Composition thus promotes maintainability.

## 4.4 INTERFACES FOR DESIGN CONSISTENCY

C# and Java provide the interface construct which promotes consistency by forcing method definition(s). A class implementing an interface must define all methods declared in that interface; otherwise, the compiler complains. Example 4.8 illustrates a simple interface: only one method prototype is declared. Every class that implements the ISocial interface must define media(). Interfaces allow different classes to provide different implementations for the same named functionality. Interfaces facilitate abstraction, like that realized by I/O streams, etc. Chapter 7 examines the design effects of interfaces and abstract classes, relative to software longevity.

**Example 4.8 C# Interface**

```
public interface ISocial  {  string   media();  }

public class chat: ISocial
{       ...
        public string media()      {     ...     }
}

public class tweet: ISocial
{       ...
        public string media()      {     ...     }
}
```

```
public class slack: ISocial
{       ...
        public string media()          {       ...      }
}
```

C# interfaces may be used to force a composing class to echo its subobject's interface. Example 4.9 redefines Example 4.7 in C# using a IWorker interface – by convention, C# interface names begin with a capital I. All methods defined in an interface must be implemented, otherwise, the compiler complains. The reworked C# class has more public methods than the C++ example because it must define transfer()to fulfill the IWorker interface. The compiler complains whenever it cannot find an implementation for a method defined in an assumed interface.

**Example 4.9  C# Echoed Interface Forced**

```
public interface IWorker
{       public double work();
        public bool   isValid();
        public int    sizeHive();
        public void   transfer();
}

public class workerB: IWorker
{       private int value;

        public workerB(int v = 100) { value = v; }

        // all four methods in IWorker
        // interface must be defined
        public double work()        {       ...      }
        public bool   isValid()     {       ...      }
        public int    sizeHive()    {       ...      }
        public void   transfer()    {       ...      }
}

public class Echo: IWorker
{       private      workerB         subObject;
        private      uint            countQueries;
        private      bool            passThru;
        ...
        public Echo(int seed = 10)
```

```
{       subObject = new workerB(seed);
        passThru = seed > 0;
}
...
// #1 immediate pass thru
public double work()
{       return subObject.work();    }

// #2 preprocess then pass thru
public bool   isValid()
{       countQueries++;
        return subObject.isValid();
}

// #3 conditional passthru
public int       sizeHive()
{       if (!passThru) return -1;
        return subObject.sizeHive();
}

// #4 pass thru then post process
public void    transfer()
{       subObject.transfer();
        countQueries = 0;
}
}
```

Composition wraps up a subobject, hiding it from the client and thus buffering the client from change and tedious management details. An interface may force full or partial exposure of a subjobject's public interface. Wrapping, with or without any exposure, is a popular design choice.

## 4.5 WRAPPERS AND DELEGATES

The terms **wrapper, delegate,** and **handle** are often used interchangeably even though meaning and intent differ. A wrapper defines and controls an interface layered over existing code, enabling code reuse. In Examples 4.7 and 4.9, the class Echo served as a wrapper. A delegate serves to provide functionality; it may be replaceable and thus provide variant behavior. In Examples 4.7 and 4.9, the class workerB served as a delegate. A handle provides access to target data without providing any functionality or augmented behavior. Chapter 3 examined smart pointers which, as handles, wrap up raw pointers to prevent memory leaks.

Wrappers are what the name implies: an extra layer or coating. This extra layer thwarts dependencies on the encapsulated code and decreases coupling, which is especially desirable when the wrapped code is unstable. Wrappers streamline the reuse of legacy code by defining a more uniform, general, or simpler interface. *A wrapper class layers an interface over an existing class* to buffer clients from change. Wrappers may redefine interfaces, modify accessibility, augment or introduce conditional restraints on delegate functionality. Though they may provide extra functionality and/ or filter requests, wrappers primarily serve to maintain a stable and usable interface.

A popular design pattern, *Façade* [Gam95], is essentially a wrapper class. Wrappers are designed using composition: the wrapper has-a subobject and wraps up the subobject interface. The implementation invariant documents what is wrapped and why, explaining any externalization via echoing or delegate replacement by the client. Is the wrapped code unstable, proprietary, or dated? What dependencies are hidden? How does the wrapper streamline use? Is the client aware of the delegate? Expectations as to improved maintainability, or performance, should be recorded. Interface invariant details of a wrapper class are similar to those of any class: describe core utility provided and restrictions on use. Wrappers promote the reuse of existing classes for new or modified applications by providing a shell to encompass an existing class. If used well, wrappers reduce software complexity and promote maintainability.

Consider the task of tracking relative data values within a collection. Reuse of a previously defined container type is expedient – modern programming languages routinely provide generic containers. A wrapper class could easily track values (such as minimum, maximum, mean) while using an encapsulated container to store data passed in by the client. The wrapper could echo as much of the embedded container's interface as required, and provide customized functionality (such as `getPrime()`, etc.). Software development is accelerated when the class designer does not have to design, implement, debug, test or verify any of the standard operations associated with a generic container.

A **delegate** class provides functionality and promotes flexibility. Delegates held as subobjects may be instantiated internally or externally. When instantiated internally, the delegate may remain anonymous to the client. When instantiated externally, the delegate must be passed in by the client and so its existence is published. Delegates exemplify effective code reuse. Example 4.10 shows an autoPay class that uses an accountVerifier object as a delegate in a composition relationship. The delegate supports a public interface that provides validation of an account; autoPay (re)uses this functionality to verify its account data members. All calls for verification are forwarded to the delegate. Consequently, autoPay does not have to redesign validation if the process for verification changes. Moreover, client code is stable as long as the accountVerifier type is retained as the delegate. The autoPay class does 'wrap up' an accountVerifier object as a delegate but replacement exposes the delegate type. Hence, whether or not autoPay is considered to be a wrapper class depends on a strict interpretation of wrapping as hiding the delegate.

**Example 4.10 C# Delegate Class**

```csharp
public class account { … }

public class accountVerifier
{       public bool approve(account a, uint amount)
        { … }
        public bool isValid(account a)              { … }
}

public class autoPay
{       private     accountVerifier     myDelegate;
        private     account             primary;
        private     account             secondary;
        …
        public      void
        replaceDelegate(accountVerifier newImproved)
        {    myDelegate = newImproved;   }

        public bool fundsAvail(uint cash)
        {    if (myDelegate.approve(primary,cash))
                        return true;
             if (myDelegate.approve(secondary,cash))
                        return true;
             return false;
        }
}
```

```
public bool upDatePrimary(account p)
{    if (!myDelegate.isValid(p)) return false;

     if (myDelegate.isValid(primary))
          secondary = primary;
     primary = p;
     return true;
}
}
```

How do delegate and wrapper differ? A delegate is a data member or passed parameter that provides functionality. A wrapper encapsulates and reuses existing code to isolate the client from change or to streamline an interface. *A delegate serves to provide utility. A wrapper serves to redefine an interface.* The two overlap. A wrapper may wrap up a delegate. Contractual design should specify any client responsibility for seeding or replacing a delegate. The implementation invariant should identify expected reuse of functionality via delegation as well as restrictions on delegate type or interface. As with any composition design, documentation should also describe cardinality, ownership, and stability (replacement) details.

Wrapper and delegate designs focus on reusing functionality. Handles control access. Why is a smart pointer considered a handle rather than a delegate? *The smart pointer safeguards memory but provides no other functionality,* no augmented, critical, type-dependent functionality. Delegates provide functionality to an enclosing class but isolate the client code from changes in service. The enclosing class can easily modify or replace its `delegate`. Though C# offers a specific delegate construct (which is easier to use than, say, C++ function pointers), the definition and use of delegates is a general design technique. For flexibility, delegates, or actually any resource, may be 'injected' into an object. We examine next this design option and the effects of externalizing what could otherwise be a fully encapsulated data member.

## 4.6 DEPENDENCY INJECTION

Software design values modularity, achieved through functional and structural (OO) decomposition, because modularity promotes code reuse, readability, and maintainability. Decomposition yields partition boundaries, also known as "seams". Shared data passes across these seams – parameters, return values, handles to static (or global) data, and other resources. With this hierarchical decomposition of code and data, dependencies are unavoidable but should be identified and constrained.

I/O is a common dependency; data sources and sinks must be identified. Novice programmers often tie input to the keyboard and output to the console. However, maintainable code relies on the abstraction of I/O streams so that input sources and output targets may vary. The utility of a type that draws data from the keyboard or sends output to the console is limited. Is encapsulating an input filename an improvement? Selection of data source and storage is still limited and changing an encapsulated filename may be expensive. When an input or output filename is not encapsulated, but passed into, say, the constructor, the class loses control over the specification of the filename and should add checks for validity. But variety is gained by parameter passing – many different filenames can be passed. Abstraction promotes flexibility – viewing data source and sink as input and output streams, rather than specific resources, releases code from dependency on a specific environment.

A filename is a dependency. Passing a filename into a method transmits a dependency from an external (client) to an internal (service) and thus illustrates **Dependency Injection**. We examine three forms of Dependency Injection, noting when memory management responsibilities impact design. Each approach alters the implicit control of composition through the external provision of a data resource (dependency). An object's internal instantiation of a delegate or resource is replaced by a dependency passed as a parameter. Any assumption that a passed ('injected') dependency is valid must be documented, preferably contractually. Even with such a contractual requirement, for safety, the class designer should design appropriate error responses to the injection of invalid dependencies. The three types of Dependency Injection – constructor, property, and method – all shift responsibility to the client. We expand this discussion in Chapter 7 to illustrate built-in support for a wide variety of encapsulated objects.

## 4.6.1 Constructor Injection

As the name implies, Constructor Injection accepts a dependency via a parameter in the constructor call. Example 4.11 displays a `Warrior` class with a constructor that (conditionally) accepts a `Weapon`. Constructor injection is preferred for a class-wide dependency with an expected lifetime association. The dependency is initialized (constructed) by assignment from a passed parameter. It is thus set once, is immediately available for all object methods, and is stable. Client responsibility is limited to point of object instantiation; clients should recognize the release of ownership.

In all forms of Dependency Injection, the class cedes complete control. A client may pass in a null or invalid handle, or an exhausted resource that cannot be reset. Class design must define appropriate error responses. Since constructors do not return a value, an error code cannot indicate rejection of the injected dependency is inadequate. Options for error handling within a constructor include: throwing an exception, reverting to a default, modifying the passed resource (if possible), or setting state to indicate compromised functionality. This last option is employed in the Warrior class which sets the armed control flag to false if the passed Weapon is a null handle or is in an invalid state. All methods that use a Weapon must then check this flag.

A key benefit of constructor injection is stable object-level variability. The client may pass any resource that satisfies the interface specified in the constructor parameter list. Since the constructor is called exactly once, any dependency so fulfilled will be set once, any error response will be executed at most once, and the resource will be available for the object's lifetime.

### Example 4.11 C# Constructor Injection

```csharp
public class Weapon
{       // ...
        public bool damaged() { ... }
}


public class Warrior
{       private      readonly  Weapon       handle;
        private      bool                   armed = true;
        // client must release Weapon ownership
        public Warrior(Weapon w)
        {   if (w == null || w.damaged())
                    armed = false;
            handle = w;
        }

        // replace defective or missing
        public bool reArm(Weapon w)
        {       if (armed && !handle.damaged())
                        return false;
                if (w == null || w.damaged())
                    return false;
```

```
        handle = w;
        armed = true;
        return true;
    }
}
```

Languages differences impact the implementation of Dependency Injection. If Example 4.11 were written in C++, the `Warrior` class must manage deallocation and explicitly determine copy semantics. Regardless of implementation language, external instantiation of a dependency in an invalid state must be addressed. The possibility of a null handle requires the addition of existence checks to all methods that use the dependency.

Unused resources yield overhead that is not easily justified. Therefore, Constructor Injection should not be used for optional dependencies without the provision for a default handle setting of null. Constructor Injection may also complicate class hierarchies, as will be seen in Chapter 6. Constructor injection though does permit test code to circumvent the use of resources that are expensive, shared, or hard to instantiate in a test harness. Test code may instantiate, and pass, a mock object into a constructor in order to syntactically satisfy a dependency that is not directly tested.

### 4.6.2 Property Injection

Property (Setter) Injection supports deferred instantiation as well as replaceable and optional dependencies. Example 4.12 shows a C# class that implements the `ISocial` interface and encapsulates a data member, `msgDelegate`, that satisfies this interface. Property Injection in this example is implemented using C# properties. The set method is conditional, rejecting dependency replacement if the number of replacements exceed a bound established in the constructor. Use of properties is not necessary; one could simply use a (conditional) set method. Actually the name of the method that sets the property does not matter in either language.

Projection Injection may be used in conjunction with Constructor Injection for postponed instantiation: the constructor sets the handle to null, if in-field, or default initialization is not used; subsequently, a mutator (setter) is invoked that accepts (and checks) a dependency. The client may set the dependency at any time but should do so before invoking any class method that uses it. Hence, Property Injection is somewhat risky: there is no guarantee that the client will initialize the dependency prior to its use. If multiple clients use the same object, then it may be unclear who is responsible for marshalling and passing the required resource. The class designer should add existence checks before the use of required injected dependencies; the cost of such conditional evaluation will be proportional to the use of the dependency.

Property Injection easily supports the replacement of a dependency and, conceptually, can be invoked an unrestricted number of times (though the class can control its response to all requests). Checks may be added to control replacement, especially if excessive resets are a concern, and to verify dependencies.

### Example 4.12  Property Injection

```
public interface ISocial
{    string media();       }

public class messageBroker:  ISocial
{       private      ISocial       msgDelegate;
        private      uint          requestLimit;
        private      int           numRequests;

        public messageBroker(uint   bound)
        {       requestLimit = 2*bound;      }
        public string media()
        { return requestLimit.ToString(); }

        public ISocial MsgDelegate
        {       get    // possible to return null
                {       return msgDelegate; }
                set
                {       if (numRequests <
                requestLimit)
                            msgDelegate = value;
                }
        }

        public string nextAdvert()
        {       if (msgDelegate == null)
                    return null;
```

```
                    numRequests++;
                    if (numRequests > requestLimit)
                        return null;

                    return msgDelegate.media();
            }
    }
```

Property Injection may be used alongside Constructor Injection to support replacement. If used by itself, Property Injection potentially reduces the strength of the dependency since a lifetime association is not required. The client may defer instantiation and/or replace the dependency. This shift of responsibility to the client is a design risk. To ensure safety, the class designer must define appropriate error responses, possibly adding exception handling code, if a needed property is not injected. Checks for existence should precede all use of the dependency. And, again, in C++, the class designer must address responsibilities associated with resource management.

Property Injection allows a wide variety of objects, in various states, to be passed in order to satisfy the dependency. In contrast, internal construction is strictly limited and Constructor Injection by itself permits variation only in initial fulfillment.

### 4.6.3 Method Injection

Method (Interface) Injection is preferred when dependencies are confined to specific method(s) and are not class-wide. If these method(s) are not invoked, there is no need to set the dependency. Method Injection suggests an efficient design, assuming methods with dependencies are few and are infrequently invoked. Since the client must fulfill the dependency for all such calls, the client may pass any dependency that satisfies the interface. A different dependency may be passed with each call, potentially providing much variety. Method Injection can be viewed as a special case of the Strategy Pattern: an object (delegate, function pointer) is passed into method and used to perform some action. By passing different objects on different calls, the action is executed differently.

An intuitive example of Method Injection is a BST (Binary Search Tree) inorder traversal that accepts, as a parameter, a container to store the sorted data. No other BST functionality (insert, delete, clear, isEmpty) requires an auxiliary container. Method Injection supports isolated and volatile dependencies. Example 4.13 illustrates a C# class that uses a `Registry` to verify current status of a data member. A `Registry` is not needed in other methods so no `Registry` subobject is retained.

**Example 4.13  C# Method Injection**

```
public class guardedClient
{     private     string     name;
      private     bool       closed;
      private     double     balance;
      // ...
      public double getBalance(Registry r)
      {  if (closed || !r.valid(name))   return 0.0;
         else                            return balance;
      }
}
```

Method Injection localizes a dependency, confining its scope to the method. A dependency so passed may be viewed as transient (i.e. a temporary resource), used one-time and then released upon exiting scope. There is little responsibility for managing a single transaction resource. Unless the resource is retained after exiting scope, a C++ class need not supply a destructor or define copy semantics for that dependency. Independent of language, mutual dependencies may be difficult to manage with Method Injection.

Constructor and Property Injection may be used together in a class design: a dependency is set in the constructor but may be replaced in a setter. Method injection is typically used on its own for a particular resource. In Example 4.11, a `Weapon` is injected via the constructor and also via a setter (`rearm()`) for replacement. Whereas, the resource `Registry` is injected only through the method `getBalance()` that uses it. All forms of Dependency Injection must account for an invalid parameters, responding with default settings, error codes, modification or exception handling. Any assumption that an injected dependency is valid must be documented, preferably contractually.

### 4.6.4  Dependency Injection Costs and Benefits

Dependency Injection transfers responsibility to the client, creating an implicit vulnerability. The client may not track state or correctly marshal resources. Efficiency may be compromised if many clients repeatedly have

TABLE 4.1    Dependency Injection Designs and Effects

| | Client Calls | Scope | Lifetime | Error Handling | Benefit |
|---|---|---|---|---|---|
| **Constructor** | Instantiation one call | Class-wide | Stable | Exceptions Defaults | Centralized |
| **Property** | Unlimited | Variable | Volatile | Existence Checks | Defer Instantiation Replacement |
| **Method** | Unlimited | Localized | None | Error Codes Exceptions | Confined use Wide Flexibility |

to acquire the same resources. Remember that *externalization of a dependency replaces a centralized acquisition of a resource with a distributed endeavor*. If the flexibility of such dispersed responsibility is warranted, then Dependency Injection is a valid design option. If not, then the overhead placed on the client, and the possible redundant resource acquisition are not justified. Table 4.1 summarizes the intent and effect of each form of Dependency Injection.

Constructor injection is a stable design since the dependency is set upon object instantiation. With appropriate error mitigation, the resource may be used immediately by all member functions. Method injection is secure and may be efficient since it restricts exposure of a dependency. If a class design has many methods that used the same dependency, each acquiring the same resource through Method Injection, then Constructor Injection may be a better design – unless the dependency is fulfilled differently in each method call. On its own, Property Injection is an insecure design since correct execution relies on the client to invoke the setter prior to any use of the dependency. Existence checks must be added to all methods that use a dependency that is not set in the constructor.

Example 4.5 illustrated internal replacement when the state of the subobject (passcode) was *internally* determined to be invalid (expired). This example could be rewritten to use Method injection for replacement: the client would then be responsible for tracking state and providing an appropriate replacement. Example 4.6 illustrated postponed instantiation wherein the subobject was *internally* instantiated upon demand. This example could be rewritten to use Property injection for instantiation: the client would then be responsible for tracking state in order to determine when the subobject should be instantiated.

Dependency Injection enables testing by decoupling a class from expensive, shared or unreliable resources such as databases, file systems, web services, etc. Testing modern software must be as efficient as possible and should not rely on 'slow' resources such as network connections. Testing must be safe and not alter shared resources such as a database. Tests that evaluate resource usage are not responsible for testing the resource, just the response to resource acquisition, release, etc. Mock objects may be constructed and 'injected' so that testing proceeds safely and efficiently. The basic scenario is: 1) client needs a resource; 2) tie client to IResource interface rather than a specific resource; 3) support different fulfillments of IResource that do not break client code; 4) use mock objects to satisfy IResource for testing.

Criticism of Dependency Injection arises from its cost and misapplication. Overuse is possible, especially since passing a parameter is syntactically easy. As noted in Chapter 6, in regard to inheritance, any design that is easily implemented leads to overuse. Dependency Injection increases client responsibility and may compromise readability by replacing an encapsulated, specific dependency with choice restricted only by an interface. The client must acquire knowledge of the dependency and assume responsibility for its acquisition. Externalization undermines encapsulation and may introduce a significant degree of redundancy. Instead of confining the realization of a dependency to a single class design, every client must fulfill the dependency; the dependency must be passed for every constructor call, or for a targeted setter (prior to invoking specific methods), or for every method that uses the dependency. Dependency Injection has also criticized for increasing build times and decreasing performance.

Dependencies must go somewhere. Should design internalize or externalize a resource? The costs and benefits of externalization may not be obvious. To design maintainable code, consider likelihood of change (resource and interface), testing requirements, and client expectations. With Dependency Injection, a good question for class designers is 'when to stop externalizing resources'?

## 4.7 DESIGN PRINCIPLES

Composition is a design approach that structurally preserves internal control while supporting variable cardinality, lifetime, association, and ownership. Delegates support code reuse while isolating clients from change. Yet, variability and testing often warrant the acquisition of a delegate via a constructor or method. Exposure of a dependency (delegate) to the client undermines encapsulation. This design tension drives the assessment of current expectations alongside anticipation of future requirements. Composition rests on encapsulation and promotes maintainability by wrapping instability. Dependency Injection externalizes the responsibility to acquire and initialize resources (dependencies). The two designs may be used together, with precise contractual details, to yield a viable, reusable structure.

The **Dependency Inversion** principle states that high-level modules should not depend on low-level modules, that is, abstractions (type definitions) should not depend upon details (e.g. encapsulated filenames). Adherence to this principle suggests the export of internal dependencies. An assumption underlying Dependency Injection, however, is that the public interface (which would include the dependency) is stable and abstract. Satisfying an abstract interface for a variety of data input sources does not expose a specific dependency; nor does it tie the client to an unstable type. Nonetheless, Dependency Injection should be judiciously employed – all data could be injected for maximum flexibility but doing so would defeat the utility of the class construct.

## 4.8 SUMMARY

Encapsulation, appropriate OO relationships and contractual design promote maintainability and code reuse. Containment is an intuitive and straightforward design since the contained objects provide little or no functionality to the designed class. Composition reflects a higher degree of type dependency but retains internal control. Type dependency drives the choice between composition (has-a) and containment (holds-a). If an object must use functionality provided by a subobject then the object is dependent on the subobject. If subobject state affects the state of the object then type dependency exists. In both cases, composition (has-a) is the appropriate relationship. Reliance on a subobject for essential functionality suggests that the composing object is of little value without the subobject.

Contrast our signal/sensor and customer/gift card examples. Since a sensor provides core functionality, the relationship modeled must be has-a: a signal cannot operate without a sensor. However, a gift card may or may not provide functionality essential to a customer. If a retail customer shops without a gift card, a holds-a relationship is appropriate. In contrast, if a guest member must use a gift card, a has-a relationship is implied.

Composition and containment permit variation in cardinality, association, lifetime, and ownership. Contractual design should specify these details in the class or implementation invariant. How many subobjects exist? Can the number be zero? Is cardinality determined upon construction? Is number of subobjects fixed across the object's lifetime? The more stable the structure, the more likely that a has-a relationship should be modeled. Likewise, consider ownership. Who owns the subobject(s)? Is ownership permanent? Ownership implies responsibility, even with implicit deallocation, and suggests a has-a (composition) relationship. Yet, the ability to transfer ownership, either assuming or releasing ownership, does not immediately imply either has-a or holds-a. What are the conditions under which such transfers occur?

Although many terms used in design discussions overlap, their definition should be clear. Wrappers may wrap delegates and reflect the same or an altered interface. Delegates may fulfill dependencies and can be injected. Dependency Injection may support postponed instantiation as well as the replacement of delegates. Interfaces may force the echo of delegates' public functionality, etc. Consistent use of terminology and explicit documentation clarifies design. Composition design may mix and match from a variety of design options: replacement, postponed instantiation, transfer of ownership, modification of state, echoing interface, variable cardinality. That's the beauty of internal control!

## 4.9 DESIGN EXERCISE

To apply the design concepts covered in this chapter, consider the following problems. First, define a class openRange that tracks the number of given integers queried within a specified range, much like the inRange class from Chapter 1 but with the ability to provide boundary values upon request. Key differences then are the lack of any state controlling access (on/off) and the provision of getters to retrieve upper and lower bound values. Next, define a class mmmRange that tracks the number of given integers queried within a specified range, much like the inRange class from Chapter 1 with support for querying minimum, maximum, and mean values. Both these class designs rest on code reuse. Hint: consider composition where the subobject is an instance of the inRange class.

A third design problem is to reuse the feeLedger class from Chapter 2 with postponed instantiation and suppressed copying. Define a trafficStats class that tracks traffic volumes and fines associated with a specific intersection. The primary functionality is to count cars, in the four directions through the intersections. A secondary task is to record fines. A `feeLedger` delegate records fines and thus is not needed until the first fine is processed.

Finally, using Dependency Injection, define a class `cyclicSeq` that generates successive values from an encapsulated arithmetic sequence, skipping values contained in a forbidden set. Appendix B.4 provides and analyzes C# solutions.

## DESIGN INSIGHTS

*Implementation Invariants records key design decisions*

*Software Design*

Much variety achievable by varying association, ownership, and cardinality

Containment models collections with weak type dependency

Subobject not essential

Composition suggests (internalized) code reuse with type dependency

Stable association, lifetimes correlated

Postponed instantiation of subobject possible

Flexible with respect to cardinality, ownership, and association

Dependency Injection

Accept subobject from client via constructor, method(s) or setter

## CONCEPTUAL QUESTIONS

1. Why are lifetime, association, and cardinality important design details?

2. Describe the major differences between has-a and holds-a.

3. When is type dependency important in design?

4. How does modification and replacement affect the stability of subobjects?

5. Why would a class designer choose to echo all or part of a subobject interface?

6. What are the key benefits and costs of postponed instantiation?

## CHAPTER 5

# Inheritance

## CHAPTER OBJECTIVES

- Define common forms of polymorphism

- Illustrate dynamic binding

- Outline virtual function tables

- Define abstract classes

- Evaluate inheritance design approaches

## 5.1 AUTOMATE TYPE CHECKING

When is an inheritance design valuable? When is it not? Unfortunately, these questions are not routinely asked. Yet, *software designers should know when to use inheritance, and when not, especially since inheritance can be simulated with composition.* To motivate appropriate use of inheritance, consider augmenting the Icon type (from Chapter 1) to support variant movement. Objects instantiated from this new IconM class can spin, slide, or hop, with the restriction that any particular IconM object is capable of only one type of movement. A spinner cannot hop, a hopper cannot slide, etc. The type of movement associated with an IconM object is set upon construction and thereafter does not change. A slider cannot spin, not now, not ever. Example 5.1 shows the monolithic class design accommodating variant movement according to the movement 'type' of an IconM object.

IconM is monolithic (meaning formed from a huge block of stone) because it carries all data and (private) functionality needed for movement variants (spin, hop, slide). As the primary functionality of an IconM object, movement determines speed and energy consumption. For each IconM object to behave consistently, the class must track movement 'subtype'. The constructor sets the 'subtype' upon instantiation – a value that should not change during an object's lifetime. Type identification is shared between the client and class designer via the 'type' value passed to constructor. IconM::move must check 'subtype', as shown in the if/else statement in move(). Such type identification must be replicated in every method that varies response according to subtype – an approach that invites inconsistency and is not maintainable.

**Example 5.1  C++ Monolithic Class for Icon Movement**

```
class IconM
{     float    speed, energy;
      int         x, y;
      //spinner, slider or hopper
      unsigned    subtype;

      bool        clockwise;      // need for spinner
      bool        expand;         // need for spinner

      bool        vertical;       // need for slider
      int         distance;       // need for slider

      bool        visible;        // need for hopper
      int         xcoord, ycoord;

      void   spin();
      void   slide();
      void   hop();
   public:
      // constructor must set subtype
      IconM(unsigned value = 1)
```

```
// use enum for readability
{    subtype = value;
     // and then test subtype to set
     // associated fields
}
// tedious subtype checking:
// subtype drives movement
void move()
{    if (subtype == 1)          spin();
     else if (subtype == 2)     slide();
     else                       hop();
}
};
```

The IconM design is not extensible. Introduction of a new movement variant forces the incorporation of a new 'subtype' value in every existence check, in every method whose response is controlled by subtype. To define 'oscillate' as a fourth type of movement, the class designer must modify the IconM class to include the oscillate subtype value and to add data and functionality for oscillation. All methods dependent on subtype, such as move(), must be modified to test for the oscillate subtype. Example 5.2 shows this tedious, error-prone means of expanding a type system, for just one method. Enumeration literals would improve readability but the need to open up a class for modification remains a design flaw.

**Example 5.2 Tedious Type Expansion without Inheritance**

```
// ALL methods in Icon that check
// subtype must be altered
//          => ERROR PRONE software maintenance
//   subtype drives movement
void IconM::move()
{    if (subtype == 1)          spin();
     else if (subtype == 2)     slide();
     else  if (subtype == 3)    hop();
     else                       oscillate();
}
```

Example 5.1 shows that variant movement requires specialized data and functionality. Subtype is internally tracked in order to select appropriate data and functionality. 'Manual' tracking, via if/else or switch statements, is onerous and error-prone. Inheritance is an attractive alternative. Essentially, inheritance is the definition (*derivation*) of a new class using (*based* on) an existing class. Historically, C++ used the terms *base* and *derived* classes. A derived class 'inherits' everything from the base class. A derived class object has an implicit base component that the compiler allocates and initializes before entering the derived class constructor.

Java popularized the terms *parent* and *child* classes but there is no conceptual limit to inheritance. A class hierarchy can be many levels deep: a derived class can inherit from another derived class. Equivalently, a child class can inherit from another child class. An initial class definition serves as the base from which multiple descendants may be defined. For consistency, we use *base* to refer to the initial class in an inheritance hierarchy and *derived* to refer to any descendant. We use the terms *parent* and *child* when considering the relationship between an ancestor and its immediate descendant.

A redesign of `IconM` using inheritance produces a more consistent and maintainable design, as shown in Example 5.3. The base `IconParent` class defines common data: location, speed, and energy reserves. Derived classes specialize movement: clockwise or counterclockwise spinning; vertical or horizontal sliding; and, hopping. Cohesion and readability are improved by the encapsulation of specialized movement in child classes. What happens now if an `oscillate` subtype is needed? Define another child class. The `IconParent` class is not affected. The `Spinner`, `Slider`, and `Hopper` classes are not affected. `IconParent` improves maintainability, a primary benefit of inheritance. **Type extension** should not impact existing classes.

**Example 5.3 C++ Icon Class Hierarchy**

```
class IconParent
{    protected:
          float speed, energy;
          int    x, y;
     public:
          // sets base values
          IconParent (float accel)
```

```
        {   speed = accel;     }

           void move ()      { ... }
};


class Spinner: public IconParent
{     protected:
           bool            clockwise, expand;
           void  spin();
        public:
          // constructor may invoke
          // parent constructor
           Spinner(float accel):
           IconParent(accel)       { ... }
           void move()             { spin(); ... }
};


class Slider: public IconParent
{     protected:
           bool            vertical;
           int             distance;
          void      slide();
        public:
           Slider(float accel):
           IconParent(accel)       { ... }
           void move()             { slide() ;... }
};


class Hopper: public IconParent
{     protected:
           bool            visible;
           int             xcoord, ycoord;
           void            hop();
        public:
           Hopper(float accel):
           IconParent(accel)       { ... }
           void move()             { hop();   ... }
};
```

Object instantiation of a descendant class triggers the invocation of
two constructors: the parent constructor fires first, followed by the child
constructor. The compiler invokes the parent no-argument constructor

unless the descendant class constructor specifies a different parent constructor in its initializer list. Essentially, the compiler patches in a call to a parent constructor as the first statement in the child constructor. In Example 5.3, the descendants of IconParent pass up a float value to the parent constructor.

IconParent data members may be declared protected rather than private to permit access from derived classes. Parent classes may also declare utility methods as protected rather than private. Protected utility functions exemplify code reuse: all derived classes can use the parent utility function so it need be defined only once. Sibling classes access the same public and protected interfaces. Sibling objects each have their own implicit copy of the parent component, sharing only static parental data. Thus, siblings share the commonality of an is-a relationship with the same parent type but sustain no direct relationship with each other.

Designing an inheritance relation, and determining the best accessibility (protected or private), may be tricky. Protected accessibility opens access only to descendants, but the number (and development) of descendant classes is not constrained. With protected accessibility, a parent cannot restrict the actions of a descendant class. Hence, protected data is more vulnerable than private. Private accessibility restricts access to the immediate class, insuring design consistency but limiting change. A parent class may keep its data private and provide conditional, protected mutators, and accessors. If descendant access is not to be restricted that data may simply be defined as protected.

In the IconParent inheritance hierarchy, data and functionality for specialized movement are defined in the appropriate child class. Consequently, the subtype field, as defined in the monolithic IconM class of Example 5.1, is no longer needed. With inheritance, subtype is no longer 'manually' tracked; the derived class name denotes the subtype. For hopping, instantiate a Hopper object; for spinning, instantiate a Spinner object, etc. An oscillate subtype is defined simply as another descendant of IconParent: its definition does not impact any existing classes. Inheritance is thus preferred for extensibility. Yet, inheritance may not always be the optimal design. We contrast inheritance and alternative designs in Chapters 6 and 7.

## 5.2 POLYMORPHISM

Inheritance is easy to implement syntactically and does not require explicit design decisions with respect to copying and ownership of the parent component. Inheritance designs are encouraged as a simple way to reuse code.

But composition also affords code reuse – the structural design of inheritance can be mimicked with composition. *Code reuse, on its own, is not a sufficient rationale for inheritance.* A composing class can gain access to all the public data and functionality of an encapsulated object that could otherwise be a "parent". If access to protected data and functionality is desired, a wrapper class can be defined with the sole purpose of opening up the protected data and functionality. *Accessibility, on its own, is not a sufficient rationale for inheritance.* What then is so important about inheritance, the major design construct of OOD?

*The true power of inheritance is behavioral modification.* A base class with a 'unifying' interface allows descendant classes to define a range of variant behavior that conforms to this inherited interface. **Heterogeneous collections** and **substitutability**, two design touchstones of extensible code, are feasible only under such a shared interface. Functionality (behavior) can vary within class hierarchies. OOPLs support polymorphism, that is, dynamic binding of function calls so that a single object handle can provide access to varying behavior at run-time.

The Greek roots of the word "polymorphism" are: many (poly) and form (morph). Polymorphism in software then refers to a function, method, class, or type name that is associated with more than one form or implementation. What are the costs and benefits of polymorphism? How does design effectively use polymorphism? It depends on the type of polymorphism.

## 5.2.1 Overloading

Software design employs different types of polymorphism. **Overloading** occurs when a function has multiple definitions, each distinguished by the **function signature** (function name and the number, type, and order of parameters). Constructors are commonly overloaded. Though not obvious, the constructor defined in Example 5.1 yields overloaded versions for the client. If the client provides a value when instantiating an IconM object, the compiler calls the constructor that takes an unsigned. If no value is provided (as when an array of C++ IconM objects is declared), the no-argument constructor is invoked; the compiler calls the constructor that takes an unsigned and patches in the default unsigned value of '1'.

Functions for routine processing may be overloaded. Example 5.4 illustrates various reset routines. Elements of an array could be reset to: a common value such as zero; a specified (passed) integer value; a value scaled up or down by an additive or multiplicative factor.

### Example 5.4 Overloaded Functions

```
void reset()
{    for (int k=0; k < size; k++)
        A[k] = 0.0;
}
void reset(double value)
{    for (int k=0; k < size; k++)
        A[k] = value;
}
void reset(bool op, int factor)
{    if (op)
        for (int k=0; k < size; k++)
            A[k] *= factor;
    else
        for (int k=0; k < size; k++)
            A[k] += factor;
}
```

## 5.2.2 Generics

**Parametric polymorphism** refers to code that has the same structure but operates on different types of data. Swapping two values is a classic example of a function that performs the same actions regardless of data type, as shown in Example 5.5. Likewise, similar actions unfold for sorting, searching, etc. Typed versions of essentially the same function are needed in a statically typed language because the compiler must know size (and thus type) to allocate space for local variables and parameters. Yet, if the underlying data type does not affect code instructions, a placeholder could stand in for the data type; the compiler could supply an actual type when needed. That is exactly what **generic**, or templated, code does.

### Example 5.5 Consistent Behavior => C++ Generic Functions

```
void swap(int& x, int& y)
{    int  hold = x;
    x = y; y = hold;
}
void swap(float& x, float& y)
{    float  hold = x;
    x = y; y = hold;
}
```

```
template<typename     someType> // generic version
void swap(someType& x, someType& y)
{     someType  hold = x;
      x = y; y = hold;
}
```

Container are good candidates for templated code since data storage and retrieval is unaffected by type. A stack pushes, pops, tests for empty, and clears no matter the type of data it stores. A priority queue stores data in order, requiring any data type stored to support comparison; otherwise, a priority queue operates independent of type, etc. Templated code is a realization of **parametric polymorphism**: functions, methods, or classes that are defined independent of a specific type. C++ advanced the use of generics by developing and disseminating the STL (Standard Template Library). One generic version easily represents an unlimited number of implementations by using a type placeholder rather than a precise type. The compiler generates a specific implementation of the generic definition when the client designates a type. Design redundancy is thus greatly reduced though code bloat may ensue.

### 5.2.3 Subtype Polymorphism

**Subtype polymorphism** requires inheritance and relies on base class references (or pointers). Subtype polymorphism provides 'automatic' support of variant behavior: the compiler tracks subtype, not the class designer (as was the case for IconM). Function call resolution is postponed until run-time, requiring an extra layer of indirection. Method invocation is processed through a base class reference (or pointer) that holds the address of an object (subtype) that conforms to the base class interface. The compiler verifies the legality of the call (proper form and public accessibility) and generates code for an indirect rather than a direct call. The call is resolved at run-time via a subtype check of the object whose address is held in the base reference (or pointer). The run-time subtype determines what method executes.

A base class reference (or pointer) is called a polymorphic handle because it can hold the address of an object of any subtype in the class hierarchy. Methods that are bound at run-time are called **virtual** (or **dynamically bound**) methods. Virtual methods usually have multiple implementations in a class hierarchy. Calling a virtual method through a polymorphic handle may yield different results on different runs of the same software. Essentially, a method is chosen, at run-time, according to the subtype of the object whose address is held in the reference or pointer. Since, the address stored in a pointer (or reference) may change from one run to another, the subtype of the indirectly addressed object may vary.

Function invocation requires the compiler to generate the code necessary to: store and transfer data, record the program counter so that control returns to the point of call upon function termination, and issue a direct or indirect JUMP statement. In C++ and C#, function calls are statically bound by default for efficiency. The compiler translates a call to a non-virtual method into a *direct* JUMP to a specific method address. Since the address patched into the JUMP statement cannot change, there is no flexibility but there is also no run-time overhead. Virtual methods offer an alternative to static binding. The compiler translates a call to a virtual method into an *indirect* JUMP statement. With dynamic binding, method selection is postponed until run-time (for details, see Section 5.5), possibly yielding variant behavior. Subtype polymorphism provides run-time flexibility by replacing the static resolution of a method call with dynamic resolution. OOD that uses inheritance well relies on (subtype) polymorphism.

*Dynamically bound function calls provide flexibility.* A virtual method call is not tied to one address. Yet, the impact on performance can be significant. The software must absorb the run-time overhead of resolving function addresses. The inability to inline virtual methods though probably degrades performance more, as examined in the next section.

### 5.2.4 Function Inlining

Function decomposition improves readability and affords code reuse by isolating functionality. However, function calls are not free. Just like any branch statement, a function call breaks sequential processing – twice, in fact: a jump to the function and then a jump back (a return). Jump statements flush the instruction pipeline, causing the loss of low-level parallelism. An excessive number of jumps negatively impacts performance.

**Inlining** is an optimization technique that replaces a function call with the body (code) of the function. Call, return, and stack frame setup are avoided with inlining, as is flushing of the instruction pipeline. Such optimization is best left to modern compilers. Although inlining may appear to be as simple as replacing a function call with the function body, local variables, and return statements must be transformed. Small functions are less likely to have several local variables or multiple returns and are easier to inline. Compilers inline small functions because call overhead may exceed execution cost. Function inlining increases code size for improved performance. Ironically though, performance may decrease if inlining triggers excessive page faults due code bloat.

The keyword 'inline' is a suggestion to the compiler. It is unadvisable to 'manually' inline via copy & paste! **Side effects** interfere with 'manual' inlining. Common side effects include: 1) change to global or static variables, heap memory, or passed parameters; 2) control flow disruptions due to secondary function calls, event notification, or raised exceptions; 3) external effects of I/O and/or resource (database, registry, etc.) alteration. Custom replacement of a function call with code does not sustain change. Everywhere a function was manually inlined rather than called must be updated upon any change to the function definition – an error-prone and tedious endeavor. In general, inlining by design rather than by compiler reduces software readability, maintainability, and possibly, portability. *Let the compiler work for you.*

Compiler resolution of a type or a function call is called 'static' because the association between name and address does not change at run-time. Static resolution (binding) of function calls yields efficient code: since the exact function to invoke is known, the compiler may inline the function. Inlining avoids the loss of instruction-level parallelism that occurs when non-sequential code, such as a JUMP statement, is executed. This efficiency though produces less flexible code; if a function call is resolved at compile-time, function selection cannot be modified at run-time. Enthusiasm for dynamic binding rests on the flexibility of run-time selection.

Dynamically resolved calls cannot be inlined. If the compiler cannot identify which function to invoke, the compiler cannot substitute actual function code for a function call, no matter how small the function. Thus, it can be especially costly to implement virtual set and get methods.

## 5.2.5 Costs and Benefits of Polymorphism

Table 5.1 summarizes these three types of polymorphism. Overloading, generics, and subtype polymorphism all provide variation with reduced design overhead. Overloaded function bodies look different. Generic function bodies and signatures look the same because identical actions are performed on different data types. Overridden methods support the same signature but alter behavior. Overloading is merely name reuse but enhances readability. Generics reduce development cost but may yield code bloat. Subtype polymorphism supports type extensibility and

TABLE 5.1  Types of Polymorphism

| | Characteristics | Distinguished by | Sample Effect |
|---|---|---|---|
| **Overloading** | Parameters differ | Function signature | Multiple constructors |
| **Generic** | Type-less | Type placeholder | Typed versions generated by compiler |
| **Subtype** | Overridden methods | Class scope | Heterogeneous collections Polymorphic delegates |

maintainability, but increases run-time overhead and may impede optimization by preventing inlining.

## 5.3  DYNAMIC BINDING

"Insanity is doing the same thing over and over again and expecting the same results", a quote misattributed to Einstein, may be common sense but can seem to contradict software execution. Run the same code twice with different input files, different results. Okay, technically, we are cheating because the two runs are not the same – the input data differs. When an input filename is chosen at run-time, different runs of the same code may yield different results. Such variability is also achieved via dynamic function selection. If a function to execute is chosen at run-time (via dynamic binding) then the same code may yield different results from one run to another. OOPLs automate such dynamic selection via inheritance.

Example 5.6 shows a simple C++ class hierarchy, where all defined methods are statically bound by default. The C# class design is similar, as shown in Example 5.9. The base (First) class defines two public methods (whoami and simple), in addition to a constructor with default values. Its immediate descendant, (Second), redefines one of these two inherited methods, (simple), and defines a new method (expand) as well. At the third level of the class hierarchy, the Third class redefines the inherited methods (simple) and (expand), and defines another public method (grand).

**Example 5.6  C++ Class Design: Default Static Binding**

```
// C++ class design - by default,
// method calls statically bound
//        efficient but rigid
class First
```

```
{   protected:
     int          x, y;
     int    level;
     public:
     First(int a = 1, int b = 10)
     {    x = a;      y = b;      level = 1;   }

     int whoami()      { return level;          }
     int simple()      { return x + y; }
};

class Second: public First
{   public:
     Second(int a = 10, int b = 100):
            First(a,b) { level = 2; }

     int simple()    { return x * y; }
     int expand()    { return x * y * level;      }
};

class Third: public Second
{   public:
      Third(int a = 100, int b = 1000):
            Second(a,b){ level = 3;}

     int simple()     { return x*y + x + y; }
     int expand()
     { return x*y*level + x + y + level;   }
     int grand()        { return (x+y)*(x+y) * level; }
};
```

With methods overridden at each level, there are three different definitions for simple()and two different definitions for expand(). What happens when simple()is invoked? Since a base class pointer (or reference) can hold the address of a base or derived class object, the firstPtr pointer array may hold addresses of any type in the First class hierarchy while the secondPtr pointer array may hold addresses of Second and Third objects. Statements #1, #2, and #3 show the initialization of the two pointer arrays. Figure 5.1 shows sample memory layout of the typed pointer arrays and objects defined in Example 5.7.

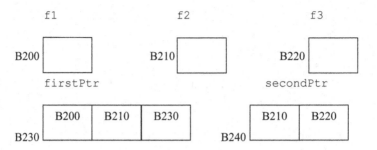

FIGURE 5.1    Sample Memory Layout for Example 5.7.

## Example 5.7  C++ Client Code: Binding Dependent on Class Design

```
int main()
// default internals, x=1, y=10
{    First        f1;
     // default internals, x=10, y=100
     Second       f2;
     // default internals, x=100, y=1000
     Third        f3;

     First*        firstPtr[3];
     Second*       secondPtr[2];

     firstPtr[0] = &f1;                         // #1
     firstPtr[1] = secondPtr[0] = &f2;          // #2
     firstPtr[2] = secondPtr[1] = &f3;          // #3

     // invocation thru C++ stack objects
     // ALWAYS statically bound
     // same as cout << First::simple(&f1)
     // #4: output 11 = 1 + 10
     cout << f1.simple() << endl;

     // same as cout << Second::simple(&f2)
     // #5: output 1000 = 10*100
     cout << f2.simple() << endl;

     // same as cout << Third::simple(&f3)
```

```
cout << f3.simple() << endl;
//#6: output 101100
//              = 100*1000+100+1000
// invocation thru C++ pointer
// can be dynamically bound
//        BUT NOT HERE!
for (int i=0; i< 3; i++)
      cout << firstPtr[i]->simple();// #7

for (int i=0; i< 2; i++)
      cout << secondPtr[i]->simple();  // #8
return 0;
}
```

The values printed at lines #4, #5, and #6 in Example 5.7 are consistent with the type of object through which simple() is invoked. The pointer type (First) in statement #7 drives the invocation of First::simple(), computing "x+y" which yields 11, 110 and 1100. Although each pointer in the firstPtr array holds the address of a different subtype, simple() does not compute "x*y" and "x*y + x +y" for a Second object and a Third object, respectively. The compiler translates a statically bound call into a direct jump statement, based on the type of the handle through which the method is invoked: a handle typed to First yields First::simple(). With such static binding, it does not matter that simple() is overridden.

The output at point #7 – 11, 110 and 1100 – differs from output at points #4–6 – 11, 1000 and 101100 – despite the source of calls being the same object: namely, firstPtr[0] contains the address of f1, firstPtr[1] contains the address of f2, etc. With static binding as the default, the method invocation firstPtr[i]->simple() is resolved at compile-time. The compiler processes statements individually so it does not track the type of the object whose address is placed in a firstPtr array pointer. The key advantage of static binding is optimization. Since simple is merely a single return statement, it can be inlined: the code to compute "x+y" is generated in place, avoiding the overhead of a function call and return.

A trace of output point #8 yields similar results: simple() always computes "x*y", yielding 1000 and 100000, despite the fact that the two pointers in the secondPtr array hold addresses of objects of different subtype. Each simple() call is statically bound, and most likely inlined, computing "x*y" upon each call. With static binding, the subtype of the object whose address is held in the pointer does not matter.

### 5.3.1 whoami() Type Identification

All three classes define a constructor with default values. Integers provided by the client (or default values) initialize protected data members x and y. Descendant classes use the initializer list to pass values to the parent constructor. First class defines a whoami() method to return the value of protected data member level (which is reset in each class constructor). whoami() identifies the subtype of any object instantiated from the First class hierarchy: 1 for a First object; 2 for a Second object; 3 for a Third object. Since the statically bound whoami() accesses level after it is reset in each descendant constructor, type is correctly identified.

whoami() is a classic identification design. Each derived class constructor must initialize the protected data member level to the unique value associated with its class definition. The client may track subtype if the association between level and its type is exposed. Design vulnerabilities include failure of the class designer to set level appropriately in constructors (especially in derived classes defined long after the original design) and failure of the client to track subtype correctly. Modern language constructs (polymorphism!) reduce the need to explicitly track type in this insecure manner. Again, the automation of type checking is a key motive for using inheritance.

### 5.3.2 Keywords for Dynamic Binding

To achieve dynamic binding in C++ and C#, one must tag a method as "virtual" in the base class. Simply adding the keyword virtual to a method name is the only change needed in Example 5.6 to trigger dynamic binding. Example 5.8 shows this modification. Once a method has been labelled virtual, all descendant classes inherit that method as a virtual method. We explain subsequently why a method is "once virtual, always virtual". Using the class First, as defined in Example 5.8, wherein simple() has been modified by virtual, the client code of Example 5.7 produces dynamic function resolution at statement #7, yielding output: 11, 1000, 101100.

**Example 5.8 C++: Virtual Keyword Tags Functions for Dynamic Binding**

```
// C++ - dynamic function identified
// by keyword virtual
// ONLY change to Example 5.6 to
// achieve dynamic binding
class First
{   protected:
      int       x, y;
      int       level;
    public:
      First(int a = 1, int b = 10)
      {    x = a;     y = b;     level = 1; }

      int whoami()            { return level; }
      virtual int simple() { return x + y; }
};

class Second: public First
{   public:
        Second(int a = 10, int b = 100):
               First(a,b){ level = 2; }

      virtual int simple() { return x*y;    }
      virtual int expand() { return x*y*level;    }
};

class Third: public Second
{   public:
        Third(int a = 100, int b = 1000):
               Second(a,b){ level = 3;  }

      virtual int simple(){ return x*y + x + y; }
      virtual int expand()
      {return x*y*level + x + y + level;}
      virtual int grand()
      {return (x+y)*(x+y)*level;   }
};
```

As in C++, default binding in C# is static; methods must be declared `virtual` in order to postpone call resolution until run-time. C# syntax though documents polymorphic behavior: to override an inherited, virtual method in C#, the descendant implementation must be labeled override. Example 5.9 illustrates this forced pairing of virtual with override. In C++, it is not necessary to label an overridden virtual method but design guidelines recommend using the keyword virtual so that class designs are self-documenting. C++11 added the optional keyword override in order to promote self-documenting code.

Using the C# First class hierarchy, defined with virtual methods, the C# client code in Example 5.10 is analogous to that of Example 5.7. At output point #G, the values printed are "1100", "100000", and then "101100". Why? The call simple() is not resolved until run-time. The compiler does not translate the method invocation into a direct JUMP statement but generates the code necessary to determine, at run-time, which method to execute. Each element of the firstPtr array is examined at run-time to determine the type of the object whose address is held therein. Since firstPtr[0] holds the address of a type First object, simple() from class First is invoked, yielding 100 + 1000. Since firstPtr[1] holds the address of a type Second object, simple() from class Second is invoked, yielding 100*1000. Since firstPtr[2] holds the address of a type Third object, simple() from class Third is invoked, yielding 100*1000 + 100 + 1000. A similar analysis confirms "1000" and then "101100" as the output at point #H.

If the keywords virtual and override were missing in the class hierarchy of Example 5.9, then the C# client code in Example 5.10 would display the same static binding effects as those from the C++ class hierarchy in Example 5.6. (Aside: a C# child class redefinition of an inherited non-virtual method produces unexpected behavior – both the child and the parent functionality runs when the non-virtual method is invoked through a child object.) What about Java? Java uses dynamic binding for all methods and does not have (or need) a keyword virtual. Hence, Java code is more consistent than C++ and C#: there is no guesswork with respect to binding; function call resolution is always postponed until run-time. Java's code consistency though is costly – the compiler cannot optimize via inlining when methods are dynamically bound.

**Example 5.9  C#: Virtual Functions Tagged and Then Overridden**

```
// C# class design - tagged
// function calls dynamically bound
```

```
//   redefined functions must be
//   labeled 'override'
public class First
{    protected int x;
     protected int y;
     protected int level;

     public  First(int p = 100, int q = 1000)
     { level = 1;          x = p;          y = q;   }

     public  virtual int simple()
     {    return x + y;  }
}

public class Second: First
{    public Second(int p = 100,
     int q = 1000): base(p,q)
     { level = 2; }

     public override int simple()
     { return x*y; }
     public virtual int expand()
     { return x*y*level; }
}

public class Third: Second
{    public  Third(int p = 100,
     int q = 1000): base(p,q)
     { level = 3; }

     public override int simple()
     {    return x*y + x + y;   }
     public override int expand()
     {    return x*y*level + x + y + level;      }

     public virtual int grand()
     { return (x+y)*(x+y)*level;
}
```

Figure 5.2 illustrates the C# object layout for the client code of Example 5.10. Since C# defines all objects as references, dynamic binding is immediately achievable through such reference variables. C# virtual

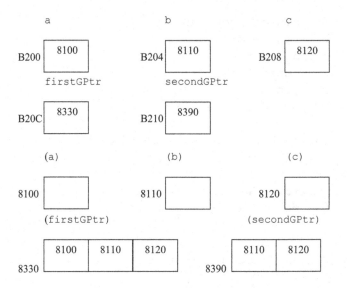

FIGURE 5.2   Sample Memory Layout for Client Code in Example 5.10.

methods are invoked in the same manner as statically bound methods – through an object (which is reference). The indirection mandated by using objects as references supports flexibility. If the address held in a reference changes, the object (and possibly the subtype) so referenced changes.

By default, C++ objects are allocated on the stack; the compiler assigns an object a (relative) address in a stack frame with space allocated for all its data members. An object allocated on the stack cannot change type due to a potential change in size. Method invocation through a C++ object cannot be postponed because the type of a stack object cannot vary. Hence, dynamic binding is not <u>directly</u> achieved through C++ objects; base class pointers must be used. A base class pointer can hold the address of a base or derived class object. Changing the address held by a base pointer may change the subtype of the object so indirectly accessed. Using C++ pointers to access (heap) objects provides the same indirection that is implicit in C#. In C++, dynamic method invocation is a two-step process: 1) declare a method virtual in the base class; and 2) call the virtual method through a base class pointer.

**Example 5.10  C# Client Code for Example 5.9**

```
// client code
First      a = new First();
```

```
Second    b = new Second();
Third     c = new Third();

First[]        firstPtr = new First[3];
Second[]       secondPtr = new Second[2];

firstPtr[0] = a;
firstPtr[1] = secondPtr[0] = b;
firstPtr[2] = secondPtr[1] = c;

for (int i=0; i< 3; i++)
   Console.WriteLine(firstPtr[i].simple());  //#G

for (int i=0; i< 2; i++)
   Console.WriteLine(secondPtr[i].simple()); //#H
```

In C++ and C#, the only method that may not be virtual is the constructor. Conceptually, a virtual constructor implies the postponement of type selection until run-time. But if type determines call resolution at run-time, how can type selection be postponed until run-time? Several design patterns mimic virtual construction [Gam95] by adding a layer to construction code. In C++, virtual methods may not be called from the constructor (restricting the application of the Template Method design pattern).

## 5.4 HETEROGENEOUS COLLECTIONS

In Examples 5.7 and 5.10, sample arrays firstPtr and secondPtr represent **heterogeneous collections**: aggregates of different (sub)types of data that can be treated uniformly. Heterogeneous collections support a common interface for all types within the collection. Varying behavior results when each different subtype satisfies the inherited interface but implements different functionality. In the First example, simple() provides variable response according to subtype.

*For software maintainability, isolate code that deals directly with type.* Code construction functions may be defined to cleanly seed heterogeneous collections. If a new subtype is added to a class hierarchy, then the construction routine(s) must be modified. Example 5.11 illustrates the isolation of object construction code in a function, and is an appropriate design for both C++ and C#. The key language difference is that C++ programmers must manage heap memory while C# programmers do not. Every C++ caller to GetObjAddr() assumes responsibility to deallocate or transfer the ownership of the heap object so received. GetObjAddr() returns the address of a default object if none of the tested conditions hold. In contrast, GetObj() returns a null reference so the caller most likely must add existence checks before using the reference.

**Example 5.11 Polymorphic Object Construction Code**

```
// object construction evaluates
// environment, possibly file input
//   generates an object of some
//   type from class hierarchy
//   => can return address of
//   any object from class hierarchy
//   => subtype of object allocate
//   determined at run-time
//
// addresses generated at run-time
//   => cannot 'guess' what
//   (sub)type of object allocated


// C++ base class pointer can
// hold address of ANY subtype object
First* GetObjAddr()
{    if (condA) return new First;
     if (condB) return new Second;
     return new Third;          // derived type II
}    // ownership of object passed back


// C# base class reference can hold

// address of ANY subtype object
First GetObj()
{    if (condA) return new First();
     if (condB) return new Second();
     if (condC) return new Third();
     return null;          // DANGEROUS
          // may yield System.NullReferenceException
}     // ownership of object passed back
```

Polymorphism promotes **extensibility** – the ability to extend a given class hierarchy without breaking existing client code. There is no need for a whoami method with well-designed polymorphism. Additional classes, say Fourth and Fifth, could easily be added to the class hierarchies in Examples 5.8 and 5.9 without affecting existing classes. These type extensions would not break any client code using polymorphic handles; the only needed change would be the object construction code isolated in GetObjAddr() or GetObj(). As shown in Example 5.12, use of a heterogeneous collection is stable since the subtype of a particular object is not exposed.

**Example 5.12 Heterogeneous Collections Hide Subtypes**

```
// initialization of heterogeneous
// collection:subtype hidden
// at compile-time, do NOT know
// type of object generated

// C++ heterogeneous collection
// => must handle heap memory
First*          bigPtrArray[100];
for (int k = 0; k < 100; k++)
    bigPtrArray[k] = GetObjAddr();

for (int k = 0; k < 100; k++)
    // dynamic behavior
    bigPtrArray[k]-> simple();
//     ...
// MEMORY MANAGEMENT: release heap
// memory before leaving scope
// deallocate dynamically allocated objects
for (int k = 0; k < 100; k++)
    delete      bigPtrArray[k];

// C# heterogeneous collection
First[]       bigArray = new First[100];
for (int k = 0; k < bigArray.Length; k++)
    bigArray[k] = GetObj();

for (int k = 0; k < bigArray.Length; k++)
    bigArray[k].simple();       // dynamic behavior
```

## 5.5 VIRTUAL FUNCTION TABLE

How does dynamic binding work? The solution is both elegant and efficient: jump tables. Each class has its own jump table, called a **virtual function table** or **vtab**, which stores the address of each virtual method inherited or redefined in the class. For dynamic call resolution, the compiler generates code to, at run-time, select a vtab according to subtype and extract the appropriate method address. *When a child class is defined, the compiler copies the parent vtab over as a default vtab for the child class.* A child class inherits a copy of its parent's vtab even if it does not add or redefine any inherited virtual methods. Hence, a method defined as virtual in the base class automatically is virtual in all descendant classes.

Each virtual method in a class definition is associated with an offset within the virtual function table (vtab). Assuming that a pointer (address) is allocated 4 bytes, the first virtual function address is stored at offset 0, the second virtual function address will be stored at offset 4, the third virtual function address will be stored at offset 8, etc. Table 5.2 displays sample vtabs for Examples 5.8 and 5.9. Since an additional virtual method was introduced in each descendant class, one additional entry was appended to each descendant class vtab. Function addresses may not be laid out exactly as illustrated: language standards need not specify layout. However, for any particular method, the offset in every vtab is the same. For example, the offset for method simple is always 0; the offset for method expand is always 4, etc. Only the addresses of methods tagged as virtual are placed

TABLE 5.2   vtabs for Examples 5.8 and 5.9

| *First* **virtual function table (vtab)** | | |
| --- | --- | --- |
| Table Entry | Virtual Function | Address (Class definition) |
| Offset 0 | simple() | First::simple() |
| *Second* **virtual function table (vtab)** | | |
| Table Entry | Virtual Function | Address (Class definition) |
| Offset 0 | simple() | Second::simple() |
| Offset 4 | expand() | Second::expand() |
| *Third* **virtual function table (vtab)** | | |
| Table Entry | Virtual Function | Address (Class definition) |
| Offset 0 | simple() | Third::simple() |
| Offset 4 | expand() | Third::expand() |
| Offset 8 | grand() | Third::grand() |

in the class vtab. Recall that `simple()`, `expand()` and `grand()`) were all denoted virtual; `whoami()` was not labeled virtual.

Whenever a child class overrides a virtual, inherited method, the corresponding entry in its vtab is updated to hold the address of the overridden implementation. If the child class does not override an inherited, virtual method, its vtab retains the parent's address. The compiler's vtab construction explains the dictum: *once virtual always virtual*. If a descendant introduces a new, virtual method, the compiler expands the vtab, adding an entry at the bottom of the table thus preserving the existing offsets.

Rather than translating a function invocation directly into to a JUMP statement, as done with statically bound calls, the compiler generates additional instructions:

1. Get type tag of the object whose address is stored in pointer/reference

2. Go to the class vtab of the resolved subtype

3. Add the offset associated with the method name to the vtab base address

4. Extract the address of the target method from the vtab entry

5. Jump to the extracted address

Since polymorphic calls are invoked through a base class pointer (or reference), the compiler *checks the base* class for a matching signature, public accessibility and the presence (or absence) of the keyword virtual. The compiler generates extra instructions to extract an address from the vtab at run-time for virtual methods. The exact vtab used, base or derived, is not selected until run-time. Thus, it does not matter if the derived class fails to override the virtual method: the derived class already has an entry in its vtab, initialized with the address of its parent's method.

## 5.6 ABSTRACT CLASSES

Heterogeneous collections are typed to an interface and can hold (addresses of) any object that conforms to that specified interface. Through this common interface, method resolution can be postponed until run-time. Data members and preliminary functionality are not typically considered part of an interface (and were not included in the C# interface construct prior to C#8.0). **Abstract classes** sit between a fully defined class and an interface. An abstract class is an incomplete type definition, providing form but not all implementation details needed for a complete definition – either public constructors or implementation of at least one method are missing. Abstract type definitions may define data members and some default functionality even if implementation is incomplete.

Example 5.13 illustrates different ways to define abstract classes. Java and C# provide a keyword abstract to indicate that a class or method is not fully defined. Any C# class that contains an abstract method must be labelled abstract. C++ does not provide such a keyword, instead using the idea of 'initializing a function to zero'. If the header of a virtual function is set "=0;", there is no definition and the method is called a **pure virtual** function. Since it has no implementation, a pure virtual function has an entry in the class vtab initialized with the value "0".

**Example 5.13  Abstract Class Definitions**

```
// C# abstract method MUST be in abstract class
public abstract class Shape
{    public   abstract void rotate(int degree);
     public   abstract void draw();
     ...
}


// C++ pure virtual methods make class abstract
class Shape
{    public:
          virtual void rotate(int) = 0;
          virtual void draw() = 0;
};


// any language: protected constructor,
// no public constructor
public class Vehicle
{    protected  Vehicle(double value) { ... }
     // public methods but no public constructor
     ...
}
```

In Example 5.13, rotate() and draw() are not implemented, so Shape is abstract. The compiler cannot support instantiation of Shape objects because it could not then resolve calls to rotate() and draw() through a Shape object. An abstract class forces inheritance. The derived class(es) must provide an implementation for each pure virtual (abstract) method inherited from the base class, or the derived class remains abstract. The descendant `Circle` class provides a definition for each inherited pure virtual function, in Example 5.14. Note that `Circle::rotate()` does nothing. Because a NOP operation defines an implementation, the client can instantiate `Circle`, as shown in Example 5.15.

An abstract class may be defined by declaring at least one method without an implementation or by not providing any public constructors. Inheritance is required to complete an abstract class definition by providing a public constructor or completing all method implementations. *A child class can invoke a protected constructor but a client cannot.* The compiler patches in a call to the protected parent constructor from the child class constructor. Thus, a client may instantiate a `Bicycle` object but not a `Vehicle`.

**Example 5.14 Derivation Completes Definition of Abstract Classes**

```
// public constructor => C# descendant
// class not abstract
public class Bicycle: Vehicle
{    public Bicycle(double value):
     base(value)   { … }
     …
}


// inherited methods defined
// => C++ descendant class not abstract
class Circle: public Shape
{        point center;
         int        radius;
      public:
         Circle(point p, int r):
         center(p), radius(r) {}
         …
         // once virtual, always virtual,
         // need not tag as virtual
         void rotate(int){}

         // for readability tag as virtual
         virtual void draw();
};
```

**Example 5.15 C++ Client Cannot Instantiate Abstract Class**

```
// cannot instantiate object from abstract class
Shape s;
// pointer can hold address of derived objects
Shape*    sptr;
// can instantiate fully defined descendant
Circle    c;

// given abstract class Shape and derived subtypes
//        Circle, Square, Triangle, Star, …
// initialize array of Shape pointers
//        each pointer can contain
//        address of different subtype
// given GetObject() to construct
// Shape subtype (on heap)
int main()
{    Shape*      composite[100];
     for (int i=0; i<100; i++)
         composite[i] = GetObjectAddr();
         …
     // what is drawn?
     for (int i=0; i<100; i++)
         composite[i]->draw();
}
```

Abstract methods are also known as **deferred methods** since definition is deferred to descendant classes. There is no code laid out in memory, and, thus, no address associated with a deferred method. The compiler cannot resolve invocation of a deferred method because, without an address, it cannot translate the function call to a JUMP statement. To forestall such errors, the compiler prevents object instantiation from an abstract class. Design intent and effect of abstract classes are summarized below.

1. **Abstract Class Design Intent:** – derived class(es) define behavior

   - Incomplete Type Definition – deferred methods

     - Function prototypes serve as placeholders

   - Polymorphism – base class defines unifying interface for class hierarchy

- Typed pointer (reference) holds address of derived object

- Calls through base pointer (reference) checked against base interface

2. **Abstract Class Design Effect:** – inheritance required

- Cannot instantiate objects

- Derived class remains abstract unless it redefines inherited deferred methods

- Heterogeneous collections expected => variant behavior supported

- Extensibility => new subtype definition does not affect existing types

An abstract class defines an interface without full implementation, establishing a dependency on descendants. The interface invariant for an abstract class should note the expected use (and any restrictions) of the public interface. The implementation invariant must describe expectations of descendant classes, the intended utility of polymorphic methods, and any default behavior that the abstract class may provide.

Abstract classes enforce a common interface for a class hierarchy, thereby promoting extensibility, and use of polymorphic objects and heterogeneous collections. Applications that handle data with common core functionality but specialized details are served well by abstract classes. For example, inventory and classification systems support common functionality (toy selection in stores, timepieces) but hold data with varying characteristics.

## 5.7 INHERITANCE DESIGNS

Examples of inheritance abound but many do not illustrate effective or maintainable designs. Budd [Bud02] classified inheritance designs and distinguished between original (clean-slate) and modified (reused) design. Clean slate designs supporting the is-a relationship are summarized in Table 5.3.

**Specialization** describes subtyping: the child class redefines or overrides core, inherited functionality. When a parent provides default behavior and a child augments or redefines that behavior, preserving the interface, the design reflects specialization. *Inheritance is anticipated but not required.*

TABLE 5.3  Budd's Inheritance Design Approaches

|  | Characteristics | Parent Interface | Relationship |
|---|---|---|---|
| Specialization | Redefines behavior | Retained | Is-a |
| Specification | Completes abstract base | Implemented | Is-a |
| Extension | Type expansion | Extended | Is-a |

For example, the Icon subtypes varied movement: a Spinner is an Icon that spins, a Slider is an Icon that slides. Specialization may not unambiguously support the is-a relationship. A priority queue is a queue that specializes enqueueing while retaining the queue interface. Yet, potential starvation could prevent the use of a priority queue in place of a regular queue, thereby undermining the is-a relationship.

**Specification** defers implementation: child classes inherit a stable interface from an abstract parent class but must implement all undefined behavior; otherwise, a child class remains abstract. No object can be instantiated from an incomplete class. Consequently, specification <u>requires</u> inheritance and expects polymorphism. *Specification differs from specialization: the child class does not refine an existing usable type but fulfills an incomplete, abstraction.* A car is-a vehicle is a clear example of specification. A vehicle interface may ensure that all derivations (car, plane, boat) move but implementation details are deferred to child classes.

**Extension** idealizes augmented functionality: child classes add new methods but do not override parent class methods. Traditionally, extension was considered a pure form of inheritance because the inherited parent interface was expanded but not compromised or redefined. In contrast, subtyping (specialization) retains the is-a relationship but its modification of inherited behavior may compromise substitutability. However, extension undermines heterogeneous collections because any method introduced by a child class is not accessible via the base class interface. Extension may not be an innocuous design, as seen in Chapter 7.

A simple example of extension is a TriAthlete class. A TriAthlete is-a BiAthlete: a TriAthlete runs and bikes like a BiAthlete but also swims. A BiAthlete is-a Runner: a BiAthlete runs like a Runner but also bikes. Extension differs from specification in that the base class is not abstract but an existing usable type. Extension differs from specialization in that the derived classes extend, and does not compromise, the inherited interface. Child classes have stricter broader interfaces. Yet, longevity is compromised when the base class interface is too narrow. For example, a heterogeneous collection typed the base Runner class may hold objects of type Runner, BiAthlete and TriAthlete. But the augmented functionality of biking and swimming is not accessible through this heterogeneous collection. Type extensibility is undermined when a client must extract (sub) type to appropriately use derived subtypes in a heterogeneous collection. Chapter 7 examines design longevity in more detail.

## 5.8 OO DESIGN PRINCIPLES

Polymorphism justifies inheritance designs. A type interface defines core functionality. A polymorphic type expects redefinition in descendant classes. Use of heterogeneous collections exercises such variant behavior. Abstract classes promote an interface that requires inheritance. The Liskov Substitutability principle tags the inherent type extensibility of inheritance and implies the power of heterogeneous collections. Any subtype can stand in for a base class object. Great variability can thus be achieved in stable software systems.

**LSP (Liskov Substitutability Principle)**
Given a type T with a subtype S defined via inheritance, any object of subtype S can serve in place of an object of type T.

Inheritance is warranted when substitutability and type checking are needed. Inheritance suggests maintainability due to the ease of type extension. Copy-and-paste reuse, a technique known to be error prone and a maintenance headache, is avoided with inheritance. In the next chapter, the choice between inheritance and composition is examined in detail.

## 5.9 SUMMARY

This chapter evaluated inheritance design, noting the benefits of substitutability and heterogeneous collections. Polymorphism provides tremendous support for the design of elegant and extensible software. Subtypes conform to the base interface while providing variant behavior. If properly used, inheritance improves software maintainability. The exposition of virtual function tables clarified dynamic binding. Abstract classes were examined as a stable design, a placeholder for extensibility and for supporting heterogeneous collections. The chapter closed by examining classic, clean slate inheritance designs that justified the overhead of this fixed, stable relationship.

## 5.10 DESIGN EXERCISES

When constructing a class hierarchy, designers must consider accessibility (protected versus private) as well as appropriate binding (default of static or dynamic via virtual designation). Child classes should reuse rather than replicate parent functionality. As the first exercise, design an inheritance hierarchy of generators where descendants specialize the retrieval of data values from an arithmetic sequence:

1. An *arithSeq* object yields the next value from its arithmetic sequence when in 'advance' mode, the previous value when in 'retreat' mode, and the current value when in 'stuck' mode. The client may alter an *arithSeq* object's mode and reset an *arithSeq* object.

2. *oscillateA* is-a *arithSeq* and thus each *oscillateA* object operates like an *arithSeq* object, except that successive values returned from an *oscillateA* object oscillate between negative and positive values.

3. *skipA* is-a *arithSeq* and thus each *skipA* object operates like an *arithSeq* object, except that values returned from a *skipA* object reflect the skipping of some number of values – this skip value should be constant but variable from object to object.

A second design exercise is to extend the *arithSeq* class hierarchy by adding a *skipA2* descendant that skips generated values that appear in a forbidden set. The client passes in the array of forbidden values (a simplified version of Dependency Injection – the injected resource is a data set rather than a database handle or network connection). Appendix B.5 provides and analyzes sample solutions.

## DESIGN INSIGHTS

Code reuse, on its own, is not a sufficient rationale for inheritance.

Accessibility, on its own, is not a sufficient rationale for inheritance.

The true power of inheritance is behavioral modification.

*Software Design*

Isolate code that is highly dependent on explicit type

**Let the compiler work for you!**

Polymorphism is powerful

Promotes type extension and thus software maintainability

Removes need for external type validation

Polymorphism is *NOT free*

Overhead of run-time binding

Prevents inlining of functions (key optimization technique)

Common interface required for heterogeneous collections

*Documentation*

Evaluate expectations for type extension

## CONCEPTUAL QUESTIONS

1. What are the three common forms of polymorphism?

2. How is the notion of type relevant in each form of polymorphism?

3. How does each type of polymorphism impact software maintainability?

4. Describe the different effects of static and dynamic binding.

5. When would a heterogeneous collection be useful?

6. What does the phrase 'once virtual, always virtual' mean?

7. Why is tracking (sub)type not desirable?

8. Define type extensibility.

## CHAPTER 6

# Inheritance versus Composition

## CHAPTER OBJECTIVES

- Identify constraints on inheritance designs
- Contrast code reuse via inheritance and composition
- Define callback and the Template Method pattern
- Illustrate the use polymorphic delegates

## 6.1 CONSTRAINED INHERITANCE

"Design first, then code": software development should start at a high level, moving from requirements to design and onto implementation. Yet, no one has a crystal ball. Current design choices may limit the reuse of legacy code, especially if language features are not analyzed. Composition may be impeded by constrained access. Inheritance may be constrained by restricted type definitions, narrow interfaces, and the overlap of binding and accessibility.

### 6.1.1 When Only Composition Is Viable

Java uses the keyword `final` to tag constants, values that cannot change. Constants are broadly used for maintainability, efficiency and readability. Arbitrary values, such as capacity limits or cardinality, are best defined as constants to confine change to only one place in the code. Constants are stored in the symbol table so that the compiler can replace them with their defined value. The symbol table may be retained for debugging but there is no program memory allocated for constants and thus no associated execution overhead for fetching a constant value from memory. Well-named constants improve readability, maintainability, and possibly performance.

Java's keyword `final` prevents change in defined functionality and type. A `final` method cannot be overridden; a `final` class cannot have descendants. C#'s keyword to tag constants (like C++) is `const` and its keyword to prevent is inheritance is `sealed`. C#'s keyword `sealed` may also be used to prevent redefinition of an overridden virtual method: a child class can thereby prevent future method modifications from descendants. Traditionally, C++ provided no keyword to prohibit inheritance. C++11 introduced the `final` modifier, to suppress inheritance. Prohibiting inheritance restricts future use and impedes testing.

### Example 6.1 Suppressed Inheritance in C#

```
public sealed class Childless
{    public bool isReady() { … }
     public int getKey()    { … }
     …
}


// failed attempt to inherit    // compilation errors

public class reuseChildless    // reuse via composition
{    private Childless myDelegate;
     …
     bool isReady()
     {    return myDelegate.isReady();    }
}

// client code: Childless obj may
// call isReady() and getKey()
// reuseChildless wrapper may call echoed isReady()
Childless       obj = new Childless();
reuseChildless wrapper = new reuseChildless();
```

Class design (in any language) may suppress inheritance. The absence of public and protected constructors prevents both clients and descendants from instantiating objects. A public static access method must then be defined in order to give the client access to class functionality. Unable to invoke a constructor to instantiate an object, a client must rely on this public static method – a design employed for resource management. See the Singleton pattern [Gam95].

Suppressed inheritance impedes testing. Recall that Dependency Injection supports testing by allowing test code to 'inject' a mock object rather than a real resource (such as a database or network connection). Mock objects, defined through inheritance, override methods dependent on external resources so that real resources are not used. Clearly, neither a C# sealed class or a C++ final class can be so easily mocked.

Whether mocked for test code or not, a sealed C# (or a final C++11) class may be reused via composition, as shown in Example 6.1. The class designer decides whether or not to echo the interface of the encapsulated object. In the reuseChildless class, isReady() is echoed; get-Key() is not. Other design variants of composition not shown in this example include postponed instantiation and delegate replacement.

## 6.1.2  When Inheritance Leaks Memory: C++ Destructors

C++ class designers must carefully manage heap memory as shown in Chapters 2 and 3. To avoid heap memory leaks and data corruption, the class designer must *explicitly determine copy semantics: suppress copying, or efficiently support deep copying.* A destructor must be defined for deallocation. The client is not responsible for invoking destructors; the compiler patches in destructor calls when stack objects go out of scope or when the delete operator is invoked for the release of heap objects.

C++ memory management becomes more complex when derived classes allocate heap memory. Destructors fire in reverse order of constructors. However, when delete is invoked through a polymorphic handle, binding affects outcome. To release a stack allocated Base class object, the compiler invokes only the Base destructor. To release a stack allocated Derived class object, the Derived destructor is invoked first, followed by the Base destructor. The Base destructor is implicitly invoked from the Derived destructor (a call to the Base destructor is the last instruction in the Derived destructor).

If a Base class pointer, myPtr, contains the address of a Base object, delete myPtr invokes the Base class destructor: no problem. What if myPtr holds the address of a Derived object? The compiler resolves calls based on the type of the pointer. With (default) static resolution of the destructor, only the Base class destructor is invoked so the Derived class destructor never runs. Failure to invoke the Derived class destructor leaks memory when the Derived class allocates heap memory. Ouch!

Example 6.2 illustrates such a hidden memory leak. Test this code yourself. When `delete` is called with `Base` class pointer b, the destructor call is statically bound so only ~`Base()` fires. Through no fault of the client, the 500 integers allocated on the heap for a `Derived` object leak because ~`Derived()` does not run. In statement #3, the client correctly calls `delete` for each heap object but with statically resolved destructors, only ~`Base()` is invoked. The client has followed convention (match every new with `delete`) and yet there is an invisible memory leak.

With heterogeneous collections, Base class pointers often hold addresses of Derived class objects. To ensure that the appropriate destructor is called through a base class pointer, destructor invocation must be postponed. The (sub)type of the object whose address in held in the base class pointer must be examined at run-time in order to execute the proper destructor(s). The setup is the same as with other virtual methods. A `Base` class pointer may hold the address of `Derived` object so if the destructor is dynamically bound, the subtype resolved at run-time determines which destructor is called first (remember invocation of ~`Derived()` automatically yields ~`Base()`).

The fix is easy: make the `Base` destructor `virtual`. That's all. No need to replicate Example 6.2. All code is the same except the keyword `virtual` is placed in front of the destructor in statement #1. Now the address of each class destructor will be placed in its vtab (because once virtual, always virtual!). Hence, for **"delete b"** (statement #2) or **"delete db[i]"**(statement #3), the compiler patches in an indirect call to the destructor via the virtual function table, postponing destructor selection until run-time. If the `Base` pointer holds the address of a `Derived` object, the `Derived` class vtab yields the address of the `Derived` destructor, so the `Derived` destructor will be called, followed by a call to the `Base` destructor. Rerun the client code of Example 6.2, once the `Base` class has been defined with a virtual destructor. No memory leaks!

**Example 6.2 Memory Leak Because Only Base Destructor Called**

```
// class hierarchy: Derived class
// allocates heap memory
class Base
{   public:

      Base()  // NO heap memory allocated
      { cout << "Base CONstructor" << endl; }
```

```
    ~Base()
    { cout << "Base DEstructor" << endl; }
};

class Derived: public Base
{    int*        ptr;
     int         size;
     ...
     // copying suppressed
   public:
     Derived(int      aSize = 900)
     {    size = aSize;
          // heap memory allocated
          ptr = new int[size];
          cout << "allocated  "
               << size << "ints" << endl;
     }

     ~Derived()              // #1
     {    delete[] ptr;   // deallocate heap memory
          cout << "deallocated  "
               << size << "ints" << endl;
     }
};

// Client code correct BUT MEMORY LEAK
void hiddenProblem()
{    Base*      b = new Derived(500);
          // Base destructor non-virtual =>
          // statically resolved
          // Compiler resolves call based on type
          // of pointer b
          // => Derived destructor not invoked:
          // 500 ints leak
     delete    b;      // #2      destructor invoked

     Base*      db[10];           // 500*10 ints leak
     for (int k = 0; k < 10; k++)
          db[k] = new Derived;

     for (int k = 0; k < 10; k++)
          delete db[k];   // #3  destructor invoked
}
```

Base class destructors must be virtual if any descendants allocate heap memory (or acquire a resource that must be released when the object goes out scope). An accepted guideline is to make the base class destructor virtual if the base class contains any virtual methods. Why? When a base class has virtual methods, it is likely that descendant objects will be used in heterogeneous collections. A virtual destructor will prevent memory leaks when a derived object is deallocated through a base class pointer. There is *no way to prevent a descendant class from allocating heap memory*. To ensure proper deallocation, tie destructor invocation to subtype rather than base type: that is, declare the base destructor `virtual`. Yet, it is not cost-effective to declare every destructor virtual. The performance cost of dynamic invocation is not the few extra instructions, which are minimal, but the inability to inline a virtual method.

What if a base class design did not anticipate that a descendant class would allocate heap memory? Heterogeneous collections may then leak memory. Opening up the base class to redefine its destructor `virtual` may not be a viable option. Although composition is often used to ameliorate interface deficiencies, wrapping cannot easily compensate for an inadequate destructor. Wrapping solves inheritance problems only <u>for descendants of the wrapper</u>. The client may have to modify its heterogeneous collections. Alternatively, with C++11, the derived class could use a container, such as a `vector`, or smart pointers in place of raw pointers to guarantee deallocation of its own heap memory while continuing to rely on statically bound base class destructor.

### 6.1.3 Inconsistent Access: C++ Accessibility and Binding

C++ allows direct suppression of an inherited interface by redefining a public inherited method as private or protected (or by redefining a protected inherited method as private). Downgrading access is called 'closing down a class' because the interface is narrowed; it is permissible in C++ but not in Java or C#. Independent of syntactic constraints, a descendant class could redefine an inherited method with no meaningful functionality (essentially a NOP). In this manner, inherited functionality could be effectively reduced, regardless of language.

The effects of binding and accessibility overlap. When the compiler resolves method invocation through a polymorphic handle, it examines *only* the base class. Accessibility (public or not) and binding (virtual or default) is checked *only* in the base class. The compiler does not evaluate descendant class modifications such as the introduction of the keyword `virtual` or (in C++) reduced access. Consider the interplay of accessibility and virtual functions in C++. The parent Diva and child Shy classes in Example 6.3 define virtual methods sing() and hum(). Both methods are public in parent Diva but child Shy reduces the accessibility of sing() by declaring it private. Hence, a client may call both sing() and hum() through a parent Diva handle but cannot call sing() through a child Shy object.

**Example 6.3 Interplay of C++ Accessibility and Binding**

```cpp
class Diva
{   public:
      virtual void sing()
        { cout << " Diva SING" << endl; }
      virtual void hum()
        { cout << " Diva HUM" << endl; }
};

// derived class suppresses part
// of inherited interface
class Shy: public Diva
{   // public inherited virtual function SUPPRESSED
    virtual void sing()
        { cout << " Shy and silent" << endl; }
  public:
    virtual void hum()  { cout << " Shy hum" << endl; }
};

// client code
Diva        b;
Shy         d;

b.sing();     // #1
b.hum();      // #2
d.hum();      // #3
d.sing();     // #4 compiler error, Shy::sing() private

Diva* bPtr = &b;
bPtr->sing();    // #5
bPtr->hum();     // #6

bPtr = &d;       // d is a Shy object; Shy::sing()
private
bPtr->sing();    // #7
bPtr->hum();     // #8
```

The sample client code in Example 6.3 compiles except for statement #4: calling sing() through a Shy object triggers an error because sing() is private in the Shy interface. Why then does statement #7 work? At that point, bPtr contains the address of a Shy object, so sing() should also be an illegal call. Yet, statement #7 compiles and produces output. To the compiler, statement #7 is no different than #5. In fact, statements #5–8 are all processed identically. Since bPtr is typed to the class Diva, the compiler verifies that sing() and hum() are defined and declared public in Diva. Since both methods are virtual, the compiler generates extra instructions for an indirect JUMP: identify (sub)type at run-time and then extract the method's address from the appropriate class vtab. The compiler generates the same indirect JUMP for statements #5 and #7 (and for #6 and #8), with the counter intuitive outcome. The private method Shy::sing() is executed at run-time because the method address for the indirect jump is extracted from the Shy vtab. A client has thus defeated a restricted interface. Dynamic binding interferes with constricted accessibility.

Tracing the compiler's actions uncovers the interplay of accessibility and binding. The compiler resolves the legality of a call relative to the type of the handle through which the call is made. If the virtual method invoked is present and public in the base class, the compiler generates the code needed for dynamic function resolution (an indirect jump). The (sub) type of the object whose address would be held in a base class pointer is not identifiable at compile-time. Subtype is extracted at run-time to identify the appropriate vtab: method accessibility is not re-checked so the suppression of an inherited, virtual method in C++ would not be evident. Put a simple output message in the class methods of Example 6.3 and test this code yourself.

When the client invokes functionality through a base class handle, the compiler resolves accessibility and binding *only* relative to the base class interface. Class designers cannot change the compiler. Hence, suppression of a public, inherited method, while possible in C++, is ineffective when virtual methods are invoked through a base class handle. A more reliable option would be to override with a NOP implementation.

## 6.2 CODE REUSE

The environmental mantra "reduce, reuse, recycle" may apply to software design as well. Reuse is clear. Recycle implies some degree of reformation (possibly through **refactoring** or wrapping). Reduce should suggest the deliberate constraint of complexity. However the class designer chooses to reuse types, the client should be minimally impacted.

Is reuse of a class more effective via inheritance or composition? Consider class minMax in Example 6.4 which accepts incoming numeric data, number by number, tracking minimum and maximum values. Say that, after the successful design, implementation, testing and deployment of minMax, a need for a minMaxMean type arises. Development could start from scratch or reuse minMax. The existing class can clearly be incorporated into a new design using either inheritance or composition. Is there a difference? *Design selection must identify tradeoffs*, as well as impact.

**Example 6.4  Simple Data Class**

```
class minMax
{           unsigned  min= MAX_INT;
            unsigned  max = 0;
      public:
            void rec(unsigned x)
            {     if (x > max)       max = x;
                  if (x < min)       min = x;
            }

            unsigned getMin()       { return min; }
            unsigned getMax()       { return max; }
};
```

Inheritance offers immediate reuse, access to protected data and functionality, and support of the is-a relationship. Composition also offers immediate reuse, but without access to protected data and functionality, and does not support the is-a relationship. Since the minMax class does not have a protected interface, or any protected data, will employing inheritance or composition to reuse minMax make much difference? Not structurally. Both the minMaxMeanInherit class and the minMaxMeanCompose class contain exactly one minMax component. The minMaxMeanInherit class has one minMax parent component and the minMaxMeanCompose class contain exactly one minMax data member. Compare the two classes, and their interfaces, as defined in Example 6.5. Is there a need for an is-a relationship?

### Example 6.5  Reuse via Inheritance vs Composition

```
class minMaxMeanInherit: public minMax
{           unsigned  sum = 0, count = 0;
     public:
           void rec(unsigned x)
           {    minMax::rec(x); // call parent method
                count++;
                sum += x;
           }

           float getMean() { return (float) sum/count;}
};
class minMaxMeanCompose
{           minMax     m;
            unsigned  sum = 0, count = 0;
     public:
           void rec(unsigned x)
           {    m.rec(x);
                count++;
                sum += x;
           }

           float getMean() {return (float) sum/count;}

           // echo subobject interface
           unsigned getMin()  { return m.getMin(); }
           unsigned getMax()  { return m.getMax(); }
};
```

With inheritance, the client automatically receives access to parent functionality via a derived object so the class designer need not echo the parent interface. When a minMax component is encapsulated as a sub-object, the client cannot directly access the public interface of the min-MaxMeanCompose object. Hence, any required public functionality of minMax must be echoed. The impact on performance is negligible when small, echoed functions, such as accessors, are inlined.

Consider reusing minMax again to define a maxRange class that tracks the difference between minimum and maximum values. Is inheritance or composition more appropriate? Again, there is no difference with respect to data: both forms of reuse require exactly one minMax component. Contrast the two classes defined in Example 6.6: they seem almost identical. Why bother with the distinction between code reuse via inheritance or composition? Echoing an interface requires more effort than simple inheritance but may be justified if any flexibility afforded by composition is used.

**Example 6.6  Reuse via Inheritance vs Composition: Full Interface**

```
class maxRangeInherit: public minMax
{  public:
     unsigned getRange() {return getMax() - getMin();}
};

class maxRangeCompose
{    minMax        m;
   public:
     unsigned getRange()
     { return m.getMax() - m.getMin(); }

     // subobject interface echoed
     void rec(unsigned x) {  m.rec(x); }
     unsigned getMin()    {  return m.getMin(); }
     unsigned getMax()    {  return m.getMax(); }
};
```

Without a designed echo or deliberate suppression, the interfaces of the maxRange classes would be different. Under the inheritance relationship, the client can extract the min and the max values via the public parent methods getMin and getMax because maxRangeInherit is-a minMax object. In contrast, with composition, the client would have no automatic access to minMax methods because minMax is a private subobject. To achieve equivalent, broad interfaces, the public interface of minMax may be echoed in the composition design, as in Example 6.6. Alternatively, to achieve comparable narrow interfaces, the inherited public methods may be suppressed (via declaration as private methods in C++) or overridden by a NOP implementation. In Example 6.7, neither class supports the public interface of the reused minMax type.

**Example 6.7  Reuse via Inheritance vs Composition: Suppressed Interface**

```
class maxRangeInherit2: public minMax
{          unsigned getMin() { }
           // suppress inherited interface
           unsigned getMax() { }
```

```
      public:
            unsigned getRange()
            { return getMax() - getMin(); }
};
class maxRangeCompose2
{          minMax m; // encapsulation hides interface
      public:
            void rec(unsigned x) { m.rec(x); }
            unsigned getRange()
            { return m.getMax() - m.getMin(); }
};
```

Choosing inheritance rather than composition, or vice versa, yields little difference in this example. Code is reused either way. There is no variability in the relationship between the minMax component and the minMaxMean type or the maxRange type. Each design defines an object with exactly one embedded minMax component. The relationship between parent and child (or object and subobject) is fixed, in terms of lifetime association, unit cardinality and ownership. There does not appear to be any need for substitutability or heterogeneous collections. Practitioners prefer composition, especially when postponed instantiation, delegate replacement, or variable cardinality are desired. Inheritance is not warranted if there is no imperative for type extensibility, substitutability, or polymorphism.

## 6.3 CLASS DESIGN: HAS-A OR IS-A?

To choose an appropriate design, evaluate tradeoffs. What are the costs and benefits of deriving a child class from a defined class versus using an instance of a defined class as a data member? The consequences of using inheritance instead of composition may not be obvious. Structurally, the layout of the two designs is similar, whether an instance of the reused class serves as a parent component or as a private data member. But design involves more than form. What is the impact on ease of use? conceptual understanding? interface flexibility? software maintainability?

Table 6.1 summarizes characteristics of composition, containment, and inheritance. With inheritance, the child class may access the public and protected interfaces of its parent class. Externally, the client may access the public interface of the parent through a child object. In contrast, composition shuts off all external access to the subobject. Internally, the composing object may access the public but not the protected interface of its subobject.

TABLE 6.1   Relationship Characteristics

|  | Association | Cardinality | Ownership | Dependency | Replacement |
|---|---|---|---|---|---|
| **Containment** | Temporary | Variable | No | No | Not relevant |
| **Composition** | Stable | Variable | Transferable | Yes | Yes |
| **Inheritance** | Permanent | Fixed: 1-1 | Implied | Yes | No |

The composing object manages the subobject and may choose to echo all or part of its interface. To avoid overhead, instantiation of the subobject might be postponed until use. With delayed instantiation, construction and cleanup responsibilities must be assumed by the class.

In comparison to inheritance, composition reduces accessibility, increases internal responsibility, and cannot directly provide the benefits of substitutability and extensibility. Yet, practitioners prefer composition over inheritance. Why? A composing object may selectively echo a subobject's interface, replace its subobject, and/or postpone instantiation of the subobject. The composing object controls its subobject while a child object may not alter its relationship with its parent component. *Inheritance is a precisely defined, implicit structural relationship that offers less flexibility than composition.* Each child object 'owns' exactly one parent component which is not replaceable or shareable. This parent component is automatically initialized via the parent no-argument constructor. The child class designer can specify the invocation of a non-default parent constructor via the initializer list but cannot circumvent the allocation or initialization of the parent component. A lifetime association exists between parent and child. The child absorbs the overhead of the parent component even if the parent is not used. With inheritance, a child object always has an implicit parent class component. This unavoidable overhead should be warranted.

Contrast minMax to IconParent from Chapter 5. IconParent defined a virtual move method that was redefined by descendant classes to provide specialized movement. A heterogeneous collection, referencing objects of any subtype defined in the IconParent class hierarchy, would support the uniform invocation of the move method. Whenever polymorphism or substitutability is required, inheritance is the best approach. *Inheritance supports type extension; composition does not.* Dynamic method selection yields maintainability; new subtypes may be added without breaking client code, as seen with IconParent. Composition designs may use polymorphic subobject(s), such a design still uses inheritance. Table 6.2 summarizes the design effects of has-a, holds-a and is-a. Composition may be preferred over inheritance because of reduced overhead.

TABLE 6.2    Relationships Effects Relative to Reused Type

| | Client Access | Internal Access | Overhead | SubObject Interface | Control |
|---|---|---|---|---|---|
| **Containment** | None | Public | Minimal | Not relevant | None |
| **Composition** | None | Public | Variable Avoidable | Suppressed May echo | Replacement Instantiation |
| **Is-a** | Public | Public Protected | Unavoidable | Support Extend | None |

In contrast to inheritance, *composition supports variable cardinality, association, ownership, and time of instantiation.* Composition retains internal control, explaining the professional preference embodied in the Composite Principle. Through composition, a class design may wrap up the interface of existing code (isolating the client from change), alter cardinality, postpone instantiation (efficiently allocating subobjects upon demand), support replacement, transfer ownership, and/or share subobjects. An immediate benefit of composition is the ability to avoid overhead when desired.

## 6.4 INHERITANCE WITH AND WITHOUT COMPOSITION

What does inheritance provide that composition does not? Type extension, substitutability, and support for heterogeneous collections. When the precise type of object needed is not known until run-time and, in fact, could easily change, as will be seen in the disassembler example in Chapter 7, polymorphism, and thus inheritance is needed. When extensibility is anticipated because code modifications may add new subtypes, as in the `IconParent` example, inheritance is again justified. When the contents of a collection may vary, and the heterogeneous subtypes contained therein offer polymorphic behavior, inheritance is required. Inheritance designs are preferred when polymorphism, substitutability, and extensibility are needed.

For example, consider a generator class that provides successive values from an arithmetic sequence. Modification of basic values so generated is desired – filtering, augmentation, replacement, etc. Inheritance designs support variability in different ways: 1) a direct is-a relationship; 2) the Template Method; and 3) composition with polymorphic delegates. See Example 6.8. The first inheritance design defines separate and stable types. The parent class defines the unifying interface for the class hierarchy and in so doing defines accessible utility via heterogeneous collections. New subtypes may be added to the class hierarchy without affecting existing types. As before, this inheritance design yields maintainable, extensible code with the expense of dynamic binding. Though the parent class establishes key utility via its defined interface, the child classes are free to override inherited methods without restriction.

**Example 6.8 C# Variant Generators**

```
// 1) inheritance alone - subtypes modify
// inherited behavior
public class Generator1
{   public  virtual int getValue() { … }
        …
}


public class filterG: Generator1
{   public  override int getValue(){ … }
        …
}


public class boostG:   Generator1
{   public  override int getValue(){ … }
        …
}


// 2) the Template Method - base class
// preserves control flow

public class Generator2
{   private int get()              { … }
    private int check(int x)       { … }
    protected virtual int process(int x) { … }
    public int getValue()
    {       // common pre-processing
            int v = get();
            // dynamic call resolved by this
            v = process(v);
            // common post-processing
            return check(v);
    }
    …
}
```

```
// 3) composition - polymorphic delegates supply
// variant behavior

public interface Igenerate      { int getValue(); }
public class modData: Igenerate
{       public virtual int getValue()
        ...
}

public class filter:            modData
{       public override int getValue(){...}
        ...
}

public class boost: modData
{       public override int getValue(){...}
        ...
}

public class Generator3: Igenerate
        // satisfied by modData delegate
{       private Igenerate   caller;
        ...

        public Generator3(Igenerate injectDelegate)
        {       caller = injectDelegate;   }

        public int getValue()
        {       return caller.getValue(); }
        ...
}
```

The second design in Example 6.8 illustrates the Template Method design pattern, an inheritance design that uses callback to retain some control over polymorphic behavior [GAM95]. The classic callback design supports variant behavior while constraining code replication. Any function may provide custom functionality alongside standard processing by 'calling back' a specialized function through a passed delegate. The passed delegate may be any object that conforms to the specified interface. In Example 6.9, chameleon is the delegate acquired via Method Injection. Typically, a test confirms that the passed delegate is not null, possibly followed by a check on the state of the delegate. Common pre-processing executes, independent of the delegate, before variant behavior is called back through the delegate. Thereafter, common post-processing runs, independent of the delegate, before the function terminates with another delegate call. The caller may pass delegates in different states or of different subtypes.

Traditionally, C++ used function pointers to implement callback. C# supports the delegate construct, which, for many, serves as a wrapped function pointer. To generalize design concepts, and to avoid a focus on language constructs, we use the term 'delegate' as defined in Chapter 4 – an object reused to provide specific functionality. With Method Injection, either C# or C++ may use (polymorphic) objects as delegates: dynamic binding and/or object state may yield different behavior via callback.

The Template Method easily supports variant behavior by partitioning functionality into three segments: common pre-processing, variant behavior, common post-processing. The common pre- or post-processing is provided through parent methods since only one version of such processing need be defined. The delegate is implicit: the this pointer. Variant behavior is 'called back' through the virtual protected process(). Essentially, the expected chain of method calls (the child calls the parent) is inverted because the parent calls (back) the child.

**Example 6.9  C++ Callback Using Polymorphic Delegate**

```cpp
// caller may pass any object satisfying
// polyDelegate interface
//    Method Injection => verify delegate
//    is not null or invalid
bool varyBehavior(polyDelegate chameleon)
{    if ( !chameleon || !chameleon->verifyID() )
          return false;
     commonPreProcessing();
     chameleon->variantSteps();
     commonPostProcessing();

     return chameleon->ok();
}
```

The Template Method design pattern, as implemented in Generator2, intrinsically uses child classes as delegates whose customized implementations of process() provide variant behavior when 'called back' in the statically bound method getValue(). The client invokes getValue() which executes a fixed sequence of actions. Stable behavior is defined by parent methods, get() and check(), which are private and may not be altered by descendants. Variant behavior arises via the virtual protected method process() which may be overridden by descendant classes: the call to process() is dynamically bound through the this reference. Descendant classes may redefine process() but may not alter structure (control flow) defined in getValue().

Delegates may be externalized, as seen in the third design of Example 6.8. In Generator3, a polymorphic handle, caller, holds the address of any type of object that satisfies the Igenerate interface. Objects from the modData hierarchy satisfy this interface and provide variety of functionality via the dynamically bound getValue(). By using a polymorphic handle, a class designer may: 1) postpone instantiation; 2) assume ownership of an external modData object; 3) replace a mod-Data object; and, 4) share a modData object. #2 and #3 suggest that the subtype of a modData object, and thus its behavior, may vary. Delegates must conform to the established interface but delegates (subtype and state) can change. In this manner, object response varies without changing the type of a Generator3 object. Dependency Injection, with the client passing in (the address of) the modData delegate, yields a maintainable design. If another subtype is added to the modData class hierarchy, no internal code change is required in the Generator3 class. In fact, with C# interfaces, a delegate instantiated from a different class or class hierarchy could be passed, as long as it conforms to the Igenerate interface.

Reliance on a polymorphic object may be built into a composition design. A composing class may have a polymorphic subobject (delegate) that fulfills key functionality. A hidden delegate may be instantiated completely internally or the client may pass in a code (int or enum) to specify the (sub)type to instantiate. Illustrated in Example 6.10, both approaches to internal construction allow the composing class to maintain control over its internal delegate and to ensure a valid initial state. The chief drawback is that extension of the delegate hierarchy requires the composing class to be opened for change. Delegate replacement likewise may be internalized.

**Example 6.10 Internal Construction of Polymorphic
Delegate in C++**

```
public class hasAdelegate
{     private delegateType  worker;

      private delegateType getDelegate(uint id)
      {    if (id == 1)     return new MinType(...);
           if (id == 2)     return new MaxType(...);
           if (id == 3)     return new MeanType(...);
           return new delegateType();
      }
      // completely internalized construction
      public hasAdelegate() {worker = getDelegate(1);
}


      // client passes in shared (code) for subtype
      public hasAdelegate(uint what)
      {    if (1 <= what && what <= 3)
                worker = getDelegate(what);
           else               worker = getDelegate(1);
      }

      public void echo()  { return  worker.echo(); }
      ...
}
```

Dependency Injection externalizes instantiation and so exposes the interface of encapsulated delegates. Through constructor, method or property injection, delegates may be passed from the client into the composite object. When delegate instantiation becomes the client's responsibility, flexibility is easily sustained: the client may supply any delegate that conforms to the specified interface; delegate subtype may change. Class design though must include appropriate error processing since the externalization of a dependency introduces the possibility of non-compliance. What is an appropriate error response if a passed delegate is null, or in an invalid state? External delegate acquisition may be less efficient than a centralized internal instantiation when multiple clients have to marshal the same resource. Vulnerability to error and client overhead (possibly evident in build times) may constrain use of Dependency Injection.

A flexible, extensible, and maintainable design may use both inheritance and composition. When a base class specifies core functionality, any object instantiated from that hierarchy provides that functionality. Composition may use a replaceable object from a class hierarchy as a delegate. By replacing a delegate of one subtype with a delegate of another subtype, internally or via Dependency Injection, a composition design alters the provided behavior without changing external type or interface.

Delegates provide functionality to an enclosing class, potentially isolating the client from change. With complete internalization, the delegate type may be modified or replaced without altering the interface of the composing class. With Dependency Injection, a polymorphic delegate may be extended without altering class internals. Again, remember the impact of language: C++ class designers must manage heap allocated (polymorphic) delegates – a task simplified through the use of smart pointers; C# class designers may use interfaces to further abstract client selection of delegates.

The choice between inheritance and composition should rest on more than code reuse. What is lost when inheritance is replaced by composition? Consider the need for extensibility, heterogeneous collections and substitutability. If neither substitutability nor heterogeneous collections is anticipated, then inheritance may not be warranted. The primary drawback of using composition rather than inheritance is the loss of direct support for polymorphism and the resulting lack of extensibility and substitutability. The variability afforded by composition with respect to lifetime, association and cardinality of subobjects increases when one considers polymorphic subobjects, as in Example 6.8. Instead of a permanent 1-1 relationship with a parent, a composing class may interact with polymorphic subobjects. The next chapter examines polymorphic delegates as a design option for the combination of two inheritance hierarchies.

## 6.5 SOFTWARE MAINTAINABILITY

A maintainable design rests on an accurate prediction of future changes but *no one has a crystal ball*. Still, an assessment of priorities – performance, control, extensibility, etc. – should guide design. Critical differences between inheritance and composition include flexibility and type support. In composition, the subobject may easily change because there is no external dependency on its hidden interface. If subobject type is changed, client code should not be impacted. With inheritance, type variability (via substitutability) and type extension are sustained by language constructs. Tracking subtype is not a client responsibility.

*Cost of change depends on interface stability.* Inheritance elevates an interface; composition hides interface. If a parent class is modified (undesirable but possible), a child class receives such updates 'automatically' through recompilation. A child class may need to make its own modifications, in response to parental changes. Any change to the base interface in a class hierarchy may significantly impact the client. If a subobject class is modified, the composing class may be forced to recompile but client code is isolated from change: if a subobject's interface changes, the composing object is not forced to modify its external interface.

*Examine design rationales carefully.* Any type relationship that reflects strong dependencies will increase coupling and decrease cohesion. While it is true that inheritance increases coupling because the child class is tightly coupled to its parent, in composition the composing class may be similarly tightly coupled to its subobject. Cohesion is decreased whenever a single type definition spans multiple classes. Inheritance decreases cohesion because the child class type definition is spread across the inheritance hierarchy. Composition also dilutes cohesion: the functionality of a composing class may be understood best by examining the subobject. *Coupling and cohesion then are not sufficient arguments for choosing either inheritance or composition.* Coupling is unavoidable due to structural and functional decomposition.

To select the most appropriate relationship, assess intended use, expected impact, and anticipated reuse. Evaluate current priorities and predict future demands. Review cardinality, ownership, and association design options as summarized below – * indicates the only option available for an inheritance design:

| | | |
|---|---|---|
| *Cardinality:* | Fixed by design* | => same for all objects |
| | Fixed at instantiation (via constructor) | => stable for object lifetime |
| | Variable during object lifetime | |
| *Ownership:* | Fixed by design* | |
| | Assumed (via Dependency Injection) | |
| | Transferable | |
| | Shareable | |
| *Association:* | Permanent* | |
| | Temporary | |
| | Stable but Replaceable | |

What are the expectations for software maintenance? If a type definition may be extended, or if heterogeneous collections must be supported, then inheritance is preferred. If a class interface is unstable, or the overhead of a parent component is significant then composition may be preferred. Software maintenance arguments must be posited carefully. *Predicting future maintenance costs is not equivalent to comparing existing overhead.*

## 6.6 OO DESIGN PRINCIPLE

Two design principles summarize this chapter's contrast of inheritance and composition designs. The **Open Closed Principle (OCP)** emphasizes type extensibility as dependent on a stable interface: *a class should be Open for extension but Closed for modification.* If OCP holds, then once a base class is deployed, its design and implementation do not change. Any number of descendant classes may be defined when the parent interface is sufficient for all descendant functionality. Essential to such design is the use of virtual functions.

The **Composite Principle** states simply that *practitioners prefer composition.* An unstated assumption is that there is no need for polymorphism, substitutability, or heterogeneous collections. As seen with `minMax`, types may be extended via composition but only effectively if there is no need for substitutability. Practitioners prefer composition because the class designer maintains internal control, more easily acquiring efficiency, and/ or flexibility. The authors of the seminal book, Design Patterns, specify a preference for object composition over inheritance despite standardizing patterns that rely on inheritance. [Gam95].

## 6.7 SUMMARY

If well defined, a class provides immediate utility and the prospect of code reuse. Both composition and inheritance reuse existing classes but may yield significant differences in overhead and maintainability. Inheritance supports the is-a relationship, yields code reuse, type familiarity, interface recognition, direct polymorphism, and substitutability. Composition supports the has-a relationship, yields code reuse, buffering of unstable interfaces, efficiency, and flexibility with respect to cardinality, association, ownership, and overhead.

Polymorphism is a key object-oriented construct. Polymorphic objects yield different behavior by postponing function resolution until run-time. Compilers use virtual function tables (vtabs) to retrieve appropriate function addresses at run-time. Polymorphic designs are flexible, promote substitutability, and support heterogeneous collections but incur the overhead of run-time binding. Software maintainability is improved by common interfaces, type extension, and "automatic" type resolution. However, *polymorphism is not free.* Extra instructions must be executed at run-time in order to support the indirect jump needed for run-time resolution. Most importantly, dynamic function calls cannot be inlined.

As a design choice, inheritance is often justified because of immediate code reuse, and the resulting reduction in development time. The child class can automatically reuse parent class functionality. But an existing class, already designed, implemented, debugged, and tested, may be reused as a parent class OR as a subobject in a composition relationship. An extended type may reuse functionality through either inheritance or composition. When considering whether to use inheritance or composition, code reuse as an argument is moot: *the class will be reused either way; it is the impact of design that must be evaluated.*

Both composition and inheritance reuse code to shorten development time. Yet, inheritance is often overused, possibly because it is trivial to define syntactically. In contrast, composition must explicitly address more design details (such as copying). For performance and maintenance, design choices should be thoroughly evaluated. Inappropriate use of inheritance may impede efficiency or require type extraction. Excessive composition may interfere with readability and maintainability. *Design should depend on goals and priorities.*

Different priorities encourage different choices for class design and relationships. OOD clearly illustrates tradeoffs and supports evaluation of both long-term and short-term cost and benefits. Enumerating contractual obligations may uncover assumptions, restrictions, and unstated obligations, possibly driving the choice between inheritance and composition. Both is-a and has-a relationships are beneficial in particular contexts. Inheritance promotes an interface, offering familiarity to a client and supporting heterogeneous collections, yielding extensibility. However, is-a is a fixed design, with unavoidable overhead. Composition wraps subobjects, and controls cardinality and instantiation. Has-a isolates client code from unstable interfaces but loses sustitutability. Interfaces somewhat mitigate this loss by preserving the ability to place objects in heterogeneous collections.

## 6.8 DESIGN EXERCISES

Three exercises aim to contrast the use of composition and inheritance. The first task is to design a `filterLedger` class, reusing the `feeLedger` class from Chapter 2. The idea is to expand the original functionality by adding an internal filter – only fees within a given range are recorded. The inRange class from Chapter 1 may also be reused to reduce development.

The second exercise starts with factorD, a type that encapsulates one non-zero value (j), provides a public divide(z) which determines if z is evenly divisible by j and, counts the number of evenly divisible values received. Next, define a second type twoFactor that reuses factorD and supports the public interface of factorD while tracking the effect of two internal values (j and k). divide(z) then determines if z is evenly divisible by j and k. For example, if j=2 and k=5, then 30 is evenly divisible by j and k but 21 is not. Reuse may be achieved through inheritance or composition.

The third exercise is to design a amorph inheritance hierarchy to model variant behavior. Each amorph is associated with a non-stable location (x, y), a size, a designated color and a brightness. dimmer is-a amorph that can vary its brightness but only if it has moved. swatch is-a amorph that can change its color but only if it has not moved more than some number of times. nimble is-a amorph that can change its size but only relative to the number of times it has moved. The final exercise is to reuse the amorph inheritance hierarchy to define shapeShifter: a type that can be switch from one amorph subtype to another after it moves. Appendix C.1 presents and discusses sample solutions.

## DESIGN INSIGHTS

*No one has a crystal ball.*

> *Software Design*
>
> Carefully evaluate the choice between inheritance and composition
>
> Assess the need for polymorphism, substitutability, and heterogeneity
>
> Evaluate any requirement for efficiency or flexibility
>
> Consider interface stability
>
> *Type Reuse*
>
> Both inheritance and composition reuse code
>
> Inheritance is a precisely defined, implicit structural relationship that offers less flexibility than composition.

Inheritance supports type extension; composition does not.

Composition supports variable cardinality, association, ownership and time of instantiation.

*Documentation*

Note type dependencies

Record rationale for design choice in the implementation invariant

Evaluate immediate use versus anticipated change

## CONCEPTUAL QUESTIONS

1. Why would suppression of inheritance be problematic?

2. Define an unstable interface and note its effects.

3. Why is the choice between is-a and has-a important?

4. When should composition be chosen in lieu of inheritance, and vice versa?

5. What is callback and why is it used?

6. How does design impact maintainability?

# III

## Effective Type Reuse

CHAPTER **7**

# Design Longevity

## CHAPTER OBJECTIVES

- Evaluate design longevity
- Illustrate effective use of abstract types
- Examine type extraction
- Assess response to inadequate interfaces
- Analyze multiple inheritance and alternatives

## 7.1 SOFTWARE EVOLUTION

A fallacy of software, as noted by Jessica Kerr, is "if it works and we don't change anything, it will keep working". Hardware upgrades, UI modifications, user base expansion (likely with different levels of proficiency), increased load, altered distribution, managed resources, etc. all affect how software 'works'. Hence, with continued use, software must evolve.

Change comes in many forms, is costly, and may conflict with prior design priorities. Performance improvements may impede portability. Interface modifications may undermine code stability. Data reclamation may impact error processing. Change may cause ripple effects, especially in tightly coupled code. Self-documenting software that conforms to requirements is easier to modify than unreadable, undocumented code. Deliberate design assesses immediate use and anticipates change. Contractual design records the rationale for design selection, identifying key assumptions that should hold even with change. Documentation of intent and effect exposes trade-offs taken – efficiency versus generality, stability versus flexibility, etc. – allowing any change agent to preserve class design.

Sustainable design rests on language constructs, as supported by the compiler, rather than on idiosyncratic customization. The compiler lays out and invokes parent and subobject constructors. The compiler controls exposure by enforcing accessibility constraints (public, private, protected). The compiler automates subtype checking by generating vtabs, and extra instructions for the indirect jumps of dynamic binding. The compiler enforces interfaces.

What characteristics of OOD *promote* longevity? A stable interface promotes maintainability. Encapsulation internalizes control and isolates clients from implementation details and internal change. Composition may wrap unstable code and manage ownership in a flexible, responsive manner. Yet, selective relaxation of encapsulation (externalization via appropriate Dependency Injection) produces a maintainable and testable design. Abstraction further enhances maintainability: clients track only interfaces.

What characteristics of OOP *sustain* longevity? Inheritance with polymorphism provides type extensibility at little or no cost to the client. With virtual methods, the definition of a new child class minimally impacts existing code. A heterogeneous collection remains stable, providing variant behavior via polymorphic calls that conform to the base class interface. We next examine a disassembler design that rests on abstract classes and supports heterogeneous collections.

## 7.2 DISASSEMBLER EXAMPLE

Sample C++ production code [Hil00] for a disassembler illustrates the utility of abstract classes. A disassembler is a reverse engineering tool, used by embedded systems engineers to quantify code coverage. As the inverse of an assembler, a disassembler translates machine code into assembly language code. Since machine code contains variable-width instructions and uses different storage types and sizes, regenerating assembly code is difficult. Disassemblers cannot regain symbolic constants and comments removed before the assembly code was converted into an executable image.

Working backwards from a trace, the process of identifying type, and thus inferring size, is one of trial and error. The disassembler must support

TABLE 7.1   Disassembler Design

| Design for Memory Location Discernment | | |
|---|---|---|
| **Disassembler** | **AbstractLocation** *class* | **AbstractLocation** *interface* |
| Collect information | Abstract type resolution | Self-identification functions |
| View trace | Multiple type interpretation | IsA(), IsReadOnly(), ... |
| Resolve R/W access | Manipulate data | Reinterpret data |
| Postpone type resolution | Re-evaluate type | Simulate different views |

multiple attempts to classify a value. Think of type identification as a guessing game. Is the value read only? Is storage a register or addressed memory? Is the location sized to hold an integer or a real? To guess, and then guess again, the disassembler must use a malleable representation for data until it can resolve storage, location, and the type of the value.

Inheritance is the most appropriate design here. A base class publishes the core functionality needed to guess storage and type. Descendant classes may modify response to reflect emerging resolution of storage and type. Since all descendants have an is-a relationship with the base, an object of a specific subtype is interchangeable with any other subtype object. Siblings may not have a direct relationship with each other but since any child object can stand in for a parent object, *all siblings are interchangeable.*

Table 7.1 lists the design intent of the abstract base class AbstractLocation shown in Example 7.1. The disassembler may use any AbstractLocation subtype, each of which represents a viable type interpretation. Summarized in Table 7.2, essential methods to guess type include: Clone() to copy an object; IsKnown() to confirm (or deny) classification of location; IsA() to confirm (or deny) ancestry; HasAddress() and IsReadOnly() to confirm (or deny) physical

TABLE 7.2   AbstractLocation Interface: Key Virtual Functions

| **Function Name** | **Purpose** | **Details** |
|---|---|---|
| Clone | Makes copy | Uses this pointer |
| IsKnown | True only if type known | Default false |
| IsReadOnly | True only if no storage | Default true |
| HasAddress | True only if needs storage | Default false |
| GetClass | Returns class enum value | Default unknown |
| IsA | Verify type test | Default: = = unknown |
| DerivesFromA | Verifies ancestry | Default: invoke IsA |
| Set | Supply value for address | Default NOP |
| GetAs | Translation between various address types | Default return 0 |

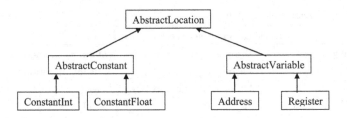

FIGURE 7.1   Disassembler Type Hierarchy

storage; and `GetClass()` to return an enumerated value, identifying type. Multiple `GetAs()` and `Set()` methods support evaluation and re-evaluation of type.

The disassembler tracks fundamental types, such as `float`, `double`, `int`, `long`, `bool`, .... For readability, an enumerated type lists the types tracked and, hence, possible interpretations of a bit string. To avoid redundant details, we consider only `float` and `int`; Figure 7.1 shows the simplified class hierarchy. Initial guesses determine whether a value is constant (cannot change) or variable (may be updated). Constants are stored in the symbol table by the compiler and retained by debuggers but do not need physical storage for execution. Memory must be assigned to variables to support changes in value.

### Example 7.1  Abstract Base Class Establishes Key Functionality

```
class AbstractLocation
{ protected:
    enum Class{unknown, constant, variable,
              constantInt, constantFloat,
              memory, register};
  public:
    virtual ~AbstractLocation() {}

    virtual AbstractLocation* Clone() const = 0;

    virtual bool IsKnown()   const {return false;}
    virtual bool IsReadOnly() const {return true;}
    virtual bool HasAddress() const {return false;}

    virtual Class GetClass() const {return unknown;}
    virtual bool IsA(Class C) const
```

```
      { return (C == unknown); }
      virtual bool DerivesFromA(Class C)
      const { return IsA(C); }

      // various Set()        - NOP as default
      // various GetAs() - zero (cast) returned as default
};
```

Attempts to classify a bit string must guess storage requirements: none for constants; physical storage for variables. Example 7.2 shows the middle tier of descendant classes which provide this distinction. AbstractConstant and AbstractVariable remain abstract because Clone() cannot be not defined without knowledge of type. AbstractVariable redefines the inherited NOP set functions as pure virtual, forcing descendant classes to define their own set functions. Why? Descendant classes gain enough type information to implement set methods.

### Example 7.2  Derived Classes Still Abstract

```
// class remains abstract because Clone() not defined
class AbstractConstant: public AbstractLocation
{ public:
      virtual ~AbstractConstant() {}

      virtual bool IsKnown() const { return true; }
      virtual Class GetClass()
      const { return constant; }
      virtual bool IsA(Class C)
      const { return (C == constant);}

      virtual bool DerivesFromA(Class C) const
      {return IsA(C)
      || AbstractLocation::DerivesFromA(C); }
};
// class remains abstract because Clone() not defined
// pure virtual set() functions force
// definition in derived class
class AbstractVariable: public AbstractLocation
{ public:
      virtual ~AbstractVariable() {}
```

```
        // variables require memory => IsReadOnly
        // overridden
        virtual bool IsReadOnly() const {return false;}
        virtual Class GetClass() const {return variable;}
        virtual bool IsA(Class C)
        const { return (C == variable);}

        virtual bool DerivesFromA(Class C) const
        {return IsA(C)
        || AbstractLocation::DerivesFromA(C); }

        // plus various set(), each "=0;"
};
```

Clone() must know type because data cannot be copied without size information. Example 7.3 shows the bottom tier of descendant classes, each of which defines Clone() so these classes are no longer abstract. Disassembler code can instantiate objects from this tier of classes, holding addresses in AbstractLocation pointers. Type interpretation is supported by polymorphic handles. If the interpretation of a bit string as a ConstantInt fails then try ConstantFloat, etc. Using a base class pointer, switching subtype is easy, just change the address held in the pointer.

**Example 7.3 Two Fully Defined Descendant Classes**

```
// class NO LONGER abstract: Clone() defined
class ConstantInt: public AbstractConstant
{      unsigned long      value;
  public:
      ConstantInt(unsigned long val): value(val) {}
      virtual ~ConstantInt() {}

      virtual AbstractLocation* Clone() const
            {      return new ConstantInt(*this); }
      virtual Class GetClass() const
            { return constantInt; }
      virtual bool IsA(Class C) const
            {      return (C == constantInt);}

      virtual bool DerivesFromA(Class C) const
```

```
        {return IsA(C)
        || AbstractConstant::DerivesFromA(C); }
        // + various GetAs() - retrieve object
        // of correct size
};
...


// class NO LONGER abstract: Clone() and set() defined
class Address: public AbstractVariable
{      unsigned long    address;
 public:
        Address(unsigned long a): address(a) {}
        virtual ~Address() {}
        virtual AbstractLocation* Clone() const
                {     return new Address(*this);    }
        virtual Class GetClass() const {return memory;}
        virtual bool HasAddress() const {return true;}
        virtual bool IsA(Class C) const
        { return (C == memory); }

        virtual bool DerivesFromA(Class C) const
        {return IsA(C)
        || AbstractVariable::DerivesFromA(C); }

        // plus various set(), each defined
};
```

## 7.2.1 Virtual Function Table

Excluding SetAs and GetAs methods, AbstractLocation has eight virtual methods; eight corresponding vtab entries are ordered consistently across subtypes so that all classes use the same offset for a specific method name. With vtab entries ordered by method declaration, IsKnown() is the third entry in all AbstractLocation hierarchy vtabs. To invoke IsKnown(), the compiler uses an offset of 8 bytes (offset of 0 for first method; offset of 4 for second method, etc.) to extract the correct address for an indirect jump from the vtab. Tables 7.3–7.5 illustrate vtabs for one branch of the class hierarchy. Overridden methods should be evident because the compiler copies the parent vtab over to the child vtab, updating each entry only when an inherited method is redefined.

TABLE 7.3   `AbstractLocation` virtual function table (vtab)

| Table Entry | Virtual Function | Address (Class definition) |
|---|---|---|
| Offset 0 | `~AbstractLocation()` | `AbstractLocation::` |
| Offset 4 | **`AbstractLocation* Clone()`** | **`0 (no valid address)`** |
| Offset 8 | `bool IsKnown() const` | `AbstractLocation::` |
| Offset C | `bool IsReadOnly()` | `AbstractLocation::` |
| Offset 10 | `bool HasAddress()` | `AbstractLocation::` |
| Offset 14 | `Class GetClass()` | `AbstractLocation::` |
| Offset 18 | `bool IsA(Class C)` | `AbstractLocation::` |
| Offset 1C | `bool DerivesFromA(Class C)` | `AbstractLocation::` |

TABLE 7.4   `AbstractConstant` virtual function table (vtab)

| Table Entry | Virtual Function | Address (Class definition) |
|---|---|---|
| Offset 0 | `~AbstractConstant()` | `AbstractConstant::` |
| Offset 4 | **`AbstractLocation* Clone()`** | **`0 (no valid address)`** |
| Offset 8 | `bool IsKnown() const` | `AbstractConstant::` |
| Offset C | `bool IsReadOnly()` | `AbstractLocation::` |
| Offset 10 | `bool HasAddress()` | `AbstractLocation::` |
| Offset 14 | `Class GetClass()` | `AbstractConstant::` |
| Offset 18 | `bool IsA(Class C)` | `AbstractConstant::` |
| Offset 1C | `bool DerivesFromA(Class C)` | `AbstractConstant::` |

TABLE 7.5   `ConstantInt` virtual function table (vtab)

| Table Entry | Virtual Function | Address (Class definition) |
|---|---|---|
| Offset 0 | `~ConstantInt()` | `ConstantInt::` |
| Offset 4 | `AbstractLocation* Clone()` | `ConstantInt::` |
| Offset 8 | `bool IsKnown() const` | `AbstractConstant::` |
| Offset C | `bool IsReadOnly()` | `AbstractLocation::` |
| Offset 10 | `bool HasAddress()` | `AbstractLocation::` |
| Offset 14 | `Class GetClass()` | `ConstantInt::` |
| Offset 18 | `bool IsA(Class C)` | `ConstantInt::` |
| Offset 1C | `bool DerivesFromA(Class C)` | `ConstantInt::` |

## 7.3 TYPE EXTRACTION

With polymorphism, the client need not 'manually' check subtypes. A method tagged as virtual in the base class triggers dynamic binding. Type extensibility is guaranteed, with little impact on client code other than the additional construction code needed for new subtype(s). Any heterogeneous collection typed to the base class interface can address objects that satisfy that interface. What if legacy code did not tag methods as virtual?

Not a problem in Java: all methods are dynamically bound. But, for efficiency, C++ and C# class designs may rely on the default of static binding. When methods in a parent class are non-virtual, the client must extract subtype to select an overridden method, as shown in Example 7.4.

Type casting converts the value of one type to the equivalent value of another type, e.g. casting integer 7 to float 7.0. Implicit casting or type conversion is performed automatically by the compiler, as when a float is assigned to an integer or vice versa. Explicit casting must be requested. Verifying type before casting is standard practice when manipulating objects from a heterogeneous collection. For type extraction, C++ provides dynamic _ cast and static _ cast; C# provides as and is. The client may use these casting operators to compensate for static binding, compromising maintainability. Client type extraction suggests poor design. To reclaim type when type is unknown, the client must test for each possibility. This tedious and error-prone type extraction requires a lengthy switch (or a multi-arm if-else) to check for all possible subtypes, a costly process of elimination.

The C++ client specifies the desired type for generic dynamic _ cast and static _ cast operators. "dynamic _ cast<Child1*> (basePtr)" attempts to reclaim the Child1 subtype from a polymorphic handle by: 1) at run-time, check the subtype of the object whose address is held in basePtr; 2a) return the address held in basePtr *if* subtype is Child1; 2b) return a zero, indicating failure, *if* subtype is NOT Child1. The statement "if (Child1* ptr = dynamic _ cast<Child1*> (HeteroDB[i]))" from Example 7.4 examines the (i+1)$^{st}$ pointer value from the heterogeneous array. If this address 'points to' a Child1 object, then the address is assigned to ptr which is of type Child1; since an address is non-zero (true), the condition is fulfilled and the rest of the if-else is skipped. If this address does NOT 'point to' a Child1 object, then zero (false) is assigned to ptr, the conditional test fails, and control flows onto the next arm of the if-else statement.

dynamic _ cast is an expensive run-time check. Using static binding for efficiency may backfire if clients must perform dynamic _ cast operations. Multiple dynamic _ cast calls cost more than a single, virtual call. static _ cast is not expensive because it does not perform a run-time check: it simply cast the given object to the specified type. static _ cast should be used only when type is known but still should be wrapped in exception handling code because an invalid cast triggers a run-time error. C#'s operators are comparable to those in C++ though the syntax is easier to read.

### Example 7.4 Type Extraction

```
// SETUP:  class hierarchy with two public
// methods in base class
// 1) void process() is NOT virtual =>
// statically bound call
//      => always yields Base functionality
// C++ descendant may simply redefine as
// 'void process()'
// C# descendant redefine (hide) as
// 'new public void process()'
// 2) void surprise() is virtual => yields
// subtype functionality

// CLIENT code uses heterogeneous collection
//     virtual function == automatic type checking
//     non-virtual function == no automatic
//     type checking
//             => manual type-checking if
//                Derived behavior desired

// C++ dynamic_cast: run-time type check
// of 'pointed to' object
// address returned if type matches;
// zero returned if cast fails
for (int i=0; i < 100; i++)
{ // virtual call, elegant: compiler sets
  // up dynamic invocation
  HeteroDB[i]->surprise();

  // clunky, tedious, not extensible
  if (Child1* ptr = dynamic_cast<Child1*>
                (HeteroDB[i]))
      ptr->process();
  else if (Child2* ptr = dynamic_cast<Child2*>
                (HeteroDB[i]))
      ptr->process();
  else if (Child3* ptr = dynamic_cast<Child3*>
                (HeteroDB[i]))
      ptr->process();
  ...           // for all relevant subtype
                // variants, test cast
```

```
      else HeteroDB[i]->process();
// catchall: unmatched subtype
}

// C# CLIENT code:  is checks type;
// as casts type when match
for (int i=0; i < 100; i++)
{  // virtual call, elegant: compiler sets up
   // dynamic invocation
   HeteroDB[i].surprise();

   // clunky, tedious, not extensible
   if (HeteroDB[i] is Child1)
   {  Child1 x = HeteroDB[i] as Child1;
      x.process();
   }
   else if (HeteroDB[i] is Child2)
   {  Child2 x = HeteroDB[i] as Child2;
      x.process();
   }
   else if (HeteroDB[i] is Child3)
   {  Child3 x = HeteroDB[i] as Child3;
      x.process();
   }
   ...            // for all relevant subtype
                  // variants, test cast
   else ...       // catchall: subtype unmatched
}
```

Without polymorphism, the client must check type in order to invoke the desired subtype behavior. In Example 7.4, process() is not dynamically bound and so must be invoked through a specific subtype. Type checking is tedious, error-prone, and not extensible. Clients must remember to check for subtypes wherever needed: a vulnerable proposition in the modern era of large-scale software. Say subtype extraction is performed in 15 different places. What happens if a new descendant extends the type hierarchy? Every place that performs type checking must add another arm to the if-else or switch statement. An update is expected in those 15 different places. What happens if one update is missed? Software becomes inconsistent and possibly unreliable. Type checking code may be isolated in a function but *the difficulty of ensuring all updates in a large software system remains.*

Testing type, incrementally, as shown in the multi-arm if statement of Example 7.4, is inefficient. If one must exclude eight subtypes before a match is found on the ninth cast, then nine dynamic casting operations are required – much more expensive than a single virtual method call. Type verification may be curtailed when the public interface of the base class provides type identification. Illustrated in Chapter 5, whoami(), defined in the base class, returned the value of data member level, as set in each descendant constructor. A single call to whoami() returns subtype identity, typically a number or an enumerated value for readability. With a known type, a static cast may directly reclaim type, as shown in Example 7.5, thereby reducing the number of casts to one. A static _ cast converts type directly while a dynamic _ cast calls wraps type casting in exception handling code, nicely returning a zero if the cast fails. Use of static _ cast is vulnerable unless the client knows the precise subtype to be reclaimed.

Although whoami() reduces the cost of type extraction, responsibility for proper use still resides externally. The client must correctly associate an ordinal value with a subtype. Hence, for safety, static _ cast should be wrapped in exception handling code. Type-checking methods cannot compensate for poor maintainability. Updating software is difficult when code relies on type extraction. Type checking code may reflect inadequate design. *Design is not extensible when a client must serve as a compiler and extract type.*

**Example 7.5  Type Reclamation with Static Cast in C++**

```
// whoami() in class hierarchy yields identifying int
//      myObj is base class pointer, just like
//      HeteroDB[i]
int typeId     myObj->whoami();

switch (typeId)
{ case 0:
  { SubType0*  ptr = static_cast<SubType0*> (myObj);
    ptr->process();
    break;
  }
  case 1:
  { SubType1*  ptr = static_cast<SubType1*> (myObj);
    ptr->process();
    break;
  }
```

```
    ...
    case 8:
    { SubType8*   ptr = static_cast<SubType8*> (myObj);
      ptr->process();
    }
}
```

As much as possible, type checking should be left to the compiler – the most consistent, secure, and maintainable approach. Client type checking is onerous, insecure, and unmaintainable. Type checking internalized within a class provides an uncomfortable middle ground: security is not much improved but efficiency is.

## 7.4 PROBLEMATIC TYPE EXTENSION

External benefits of reuse via inheritance may be limited. Although any object of a derived class may be substituted for an object of the base class, public utility is restricted to that published in the base class interface. If a derived class extends its inherited interface by defining additional methods, one cannot call those additional methods through a base handle. In Example 7.6, the child function LOL() cannot be invoked through a parent handle. Code must be added to extract subtype in order to invoke LOL() – not a maintainable design.

A child class may expand its inherited interface. Such extension was traditionally considered a 'pure' form of inheritance. However, when heterogeneous collections are typed to the base class, the client must extract (sub)type to invoke any method not in the base interface. Distinguish between interface extension which may be problematic and behavior extension (redefinition) which usually is not.

**Example 7.6  Child Extends Parent Interface**

```
class   narrowParent  {..};
class   widerChild: public narrowParent
{    ...
    public:
        // method added to interface, not in parent
        // interface
        void LOL();
        ...
};
....
```

```
// client code
narrowParent*     db[100];
for (int j = 0; j < 100; j++)
      db[i] = GetObj();
// LOL() not in base interface => compilation error
for (int j = 0; j < 100; j++)
{     db[i]->process();     // ok, in parent interface
}
// must add code to extract subtype in order
// to invoke LOL()
for (int j = 0; j < 100; j++)
{     db[i]->process();     // ok, in parent interface

      if (widerChild* ptr = dynamic_cast
            <widerChild*> (db[i]))
          ptr->LOL();      // child class pointer used
}
```

In an inheritance design, defining the base class interface may be the most crucial decision. A base class interface is inadequate when a derived class: 1) wishes to override inherited methods but the base class uses static binding; 2) expands its inherited interface. In both cases, clients must extract type. How can design reduce the burden of type checking when future modifications and expansions cannot be predicted? When the base class provides type identification via a method like whoami(), as shown in Example 7.5, the scope of type checking is reduced. The client does not need an exhaustive sequence of checks to identify type because a whoami() query yields immediate type identification. We next examine a design with such internalized type reclamation.

Example 7.7 sketches a class hierarchy for Creatures defined for gameplay. Only the virtual reCharge() method is of interest. Assume that each Creature has a store of energy reserves which is tapped when reCharge() is invoked. Descendants of the base class may refine the inherited reCharge() but must match its signature. Hence, a derived class cannot accept a parameter when overriding reCharge(). What if a new subtype, predator, recharges by consuming energy reserves from another Creature? Unfortunately, the Creature interface cannot accommodate this modification.

**Example 7.7 Mask Type for General Use**

```
public class Creature
{    protected uint reserves;
     public virtual bool reCharge()              { ... }
}
public class agileCreature: Creature
{    ...
     public override bool reCharge()              { ... }
}
// method signatures do not match => cannot override
//    if method defined as virtual, new entry in vtab
//    extended interface not accessible via base
//    class interface
{    ...
     public bool reCharge                          { ... }
}
public class CreatureG
{    protected object reference = null;
     ...
     public virtual bool reCharge(object
       handle = null) { ... }
}
public class agileCreatureG: CreatureG
{         ...
     public override bool reCharge(object
       handle = null)    { ... }
}
public class Scavenger: CreatureG
{    ...
     public override bool reCharge(object c = null)
     {      // type conversion code
          CreatureG           yummy = c as CreatureG;
          if (yummy = = null) return false;
          ...
          return true;
     }
}
```

A generic placeholder leaves room for variability. In Example 7.7, class CreatureG alters the Creature class only by introducing a parameter of type object, with a default value of null, into the signature of reCharge(). Clients do not have to pass a parameter – a call to `reCharge()` without a parameter causes the compiler to insert a null reference. In C#, `object` is a generic reference that can hold the address of any type of object. C++ use the `void` pointer (`void*`) as a generic address holder. If the optional parameter is provided, as expected in the `Scavenger` class, then the class code internally extracts the type and proceeds to use the parameter as intended.

A derived class interface that is incompatible with the base interface may indicate that inheritance is a suboptimal design. Consider a tricky question, from an OOD perspective: is a square a rectangle? Mathematically, yes, a square is a rectangle where width matches length. However, a rectangle base class may support `chgLength()` and `chgWidth()` methods. To conform to the inherited interface, the square descendant class may: 1) override the inherited methods so that any change to length (or width) results in an equal adjustment to width (or length) – a change that is misleading to the client; 2) use the inherited methods as is, resulting in square objects that do not retain the property of equal width and length. What if the square class just wants to support a `resize()` method that takes a single value to modify both width and length, preserving the property of a square?

Maintainability concerns may arise even when an interface is sufficient. If a base class employs Constructor Injection then its descendants must pass the dependency up (through their constructor initializer lists). If these descendants have descendants, then the base class dependency must pass through three constructors, etc. With Constructor Injection, a dependency propagates down from the base class to all descendants. Combining interfaces may also be problematic, as seen next with multiple inheritance.

## 7.5 MULTIPLE INHERITANCE AND ITS SIMULATION

Multiple inheritance establishes two (or more) is-a relationships when a child class inherits from two (or more) parent classes, providing the usual benefit of code reuse. Multiple inheritance increases software complexity: cohesion decreases because the child class definition is spread across three or more classes; coupling increases because the child class is tied to two or more parent classes. Unlike single inheritance, it may not be clear what the child class reuses from a *specific* parent.

C# and Java do not support multiple inheritance, possibly because the costs of such complex designs often outweigh benefits. Multiple inheritance is supported in C++ but must be simulated in C# and Java. Examples of multiple inheritance that cannot be designed as well, or better, by simulating multiple inheritance are few. The classic Observer pattern [Gam95] used multiple inheritance – the need for this pattern has decreased with support for event notification in modern programming languages.

### Example 7.8 Multiple Inheritance

```
class Signal
{   protected:
        bool              on = true;
        unsigned          count = 0;
    public:
        int feedback(int value)
        {       if (!on)          return -1;
                count++;
                on = value != 9999;
                return count;
        }

        bool isOn()    {        return on;  }

        void reset()
        {       count = 0;
                on = true;
        }
};

class Counter
{       unsigned          count;
        unsigned          max;
        unsigned          min = UINT_MAX;
    public:
        unsigned getMax()         {        return max;   }
        unsigned getMin()         {        return min;   }

        void record(unsigned k)
        {       if (k < min)  min = k;
                if (k > max)  max = k;
                count++;
        }

        void reset()
        {       count = max = 0;
                min = UINT_MAX;
        }
};
```

```
class signalCounter: public Signal, public Counter
{   public:
        void reset()     // redefinition forced
        {        record(Signal::count);
                 Signal::reset();
        }
};
```

Example 7.8 defines a multiply inherited signalCounter, a type that combines a Signal (receives and counts values until an error code (9999) is received) with a Counter (receives values indefinitely, recording only the minimum and maximum values). The combined type tracks the minimum and maximum number of values processed before an error code is received. The two parent classes Signal and Counter provide distinct functionality but reset() has the same signature in both parent classes. Both methods are inherited so which method executes when invoked through a signalCounter object? When two parent classes contain the same method with identical signatures, it is unclear (ambiguous) which method is called through a child object. *Compilers cannot handle ambiguity.* Compilers follow long, complex directions, and use definitive rules when faced with a choice. Compilers cannot capriciously select an option. To remove ambiguity, the child class designer must override all methods in an overlapping interface. The signalCounter class must redefine reset() so that the compiler may resolve a call through a child object.

**Ambiguity** prevents compilation. The C++ language standard does not specify any correlation between the declaration order of multiple parents and the layout of parent components. Thus, compilers cannot prioritize parents in order to select methods from overlapping interfaces. Without a rule for resolving overlapping interfaces, the compiler generates an error. Ambiguity forces the child class to override any method inherited from two (or more) parents. The child class may call zero, one or both parent methods in its overridden method.

## 7.5.1 Design Difficulties

Diamond inheritance occurs when two parents of a child class both inherit from the same ancestor, giving a child object two ancestor components, one embedded in each parent component. Such redundancy is not easily resolved. C++ **virtual inheritance** tags possible redundancy. The class designer must anticipate diamond inheritance and derive each parent class 'virtually', for example class parent1: public virtual commonC and class parent2: public virtual commonC. When two parents inherited "virtually" from a common ancestor, the compiler suppresses the redundant copy in the multiply inherited child and adjusts internal references. Hence, class child: public parent1, public parent2 has only one commonC component. Virtual inheritance is only effective if all parents derived virtually from their common ancestor. Unfortunately, no one can accurately predict all future uses of a class and tagging every inherited relationship as 'virtual' is overkill, especially since diamond inheritance is uncommon.

## 7.5.2 Single Inheritance with Composition

Multiple inheritance may be mimicked via single inheritance with composition: an encapsulated delegate serves as the rejected parent. The delegate is considered a subordinate type given the loss of an is-a relationship. Type dependency should drive design: the stronger the association between parent and child types, the more feasible the parent's reuse via inheritance rather than composition. For example, [Din14], the design of a studentEmployee class subordinates the Employee class because the most important functionality of a studentEmployee is being a Student. When a child has comparably dependency on both parents, the choice of parent to subordinate is arbitrary. In Example 7.9, Counter is subordinated, with the assumption that the feedback functionality of Signal is more essential that identifying minima and maxima.

For full simulation, the composing class should echo the functionality of the subordinated parent. C# interfaces effectively mimic the is-a relationship and may be used to force implementation. Since C# does not support multiple inheritance, simulation must subordinate at least one parent. Use of an interface though provides design consistency and supports heterogeneous collections. Without the interface construct, a C++ class is not forced to echo any functionality, possibly yielding design omissions.

**Example 7.9  C++ Subordinated Parent**

```
class signalCounter: public Signal
{       Counter             delegate;
    public:
        signalCounter(unsigned d = 1000): Signal(d) { }
```

```
void reset()
{       delegate.record(count);
        Signal::reset();
}

// echo
unsigned getMax() { return delegate.getMax();}
unsigned getMin() { return delegate.getMin();}
void record(unsigned k) { delegate.record(k);}
};
```

### 7.5.3 Simulation without Inheritance

Inheritance may be avoided completely. Advantages of composition – transfer of ownership, postponed instantiation, polymorphic delegates, subobject replacement, variable lifetime, association, and cardinality – accrue when each parent type is subordinated to data member status. C# interfaces force key functionality to be echoed thus mitigating the loss of the is-a relationship with demoted parent(s).

Delegates may be instantiated internally or assume ownership from a caller. Internal construction retains control and obviates the need for error checking but is not extensible: defining a new delegate subtype may require altering construction of the internal delegate. In contrast, Dependency Injection requires defined error response but supports maintainability. Delegates are always held indirectly in C# because all objects are references. C++ must use base class pointers to reference delegates, in order to achieve type extensibility. With a polymorphic delegate, type extension (of the delegate subtype) does not affect class implementation since the client passes the delegate.

When the client injects a null object, via Constructor, Method, or Property Injection, an appropriate error response should be triggered. Standard error handling includes throwing an exception, returning an error code (not possible with Constructor Injection), replacing the null reference with a default instantiation, or peppering checks for null throughout the class.

signalCounter in Example 7.10 uses Signal and Counter delegates, displaying both internal and external approaches to instantiating delegates. The Signal object is internally constructed; a Counter object is injected. Any object that satisfies the CounterI interface may be passed into the signalCounter constructor. If the Counter class is extended, both signalCounter and legacy client code are stable. Variant behavior might be observed if a client passes in a new Counter subtype delegate but the internals of signalCounter are unaffected. However, class code must check that the data member externalObj is not null wherever it is used. Similar observations arise from the Property Injection of any delegate object typed to the CounterI interface. In contrast, the Signal delegate is constructed internally so there is no need for error checking. However, if a new Signal delegate subtype is defined, the signalCounter class cannot use it without altering its internal construction code.

**Example 7.10  C# Double Composition => Internal Type Extensibility**

```
// interfaces forces provision of Signal & Counter
// functionality
//      => supports heterogeneous collections
public class signalCounter: SignalI, CounterI
{   private   SignalI        internalObj;
    private   CounterI       externalObj;

    // Constructor Injection
    public signalCounter(CounterI b, uint choice)
    {   //any subtype typed to CounterI is
        //legal => extensible
        //external instantiation => client may
        //pass null object
        externalObj = b;

        //internal instantiation => not extensible
        //=> subtype selection altered for new
        //delegate subtype
        if (choice == 1)   internalObj = new signal();
        else if (choice == 2)
                internalObj = new skipSignal();
        ...
        else            internalObj = new flashSignal();
    }
```

```
// must check for existence of Counter object
public void record(uint k)
{       if (externalObj = = null) return;
        externalObj.record(k);
}


// Property Injection
public bool switchCounter(CounterI loud)
{       if (loud = = null)      return false;
        externalObj = loud;
        return true;
}
}
```

When simulating multiple inheritance, the choice between double composition or single inheritance with one subordinate parent may not be clear. If the child type sustains a strong is-a relationship with one parent type, then single composition is warranted. Otherwise, double composition yields control and flexibility. Data members held indirectly, via a base class pointer or reference, can be: 1) stubbed out until needed (postponed instantiation); 2) replaced (with the same or a different subtype); 3) internally or externally constructed; and, 4) used conditionally.

Multiple inheritance may proceed smoothly when inherited interfaces are compatible but not overlapping. Stable interfaces support maintainability and may be established via abstract classes in either language or by C# interfaces. Defining an abstract class (or a C#8 interface) does not require a definition of data members or default functionality. Even without such baseline definitions, multiple inheritance increases complexity, especially when the two 'parent' types represent class hierarchies that continue to evolve.

## 7.6 CLASS HIERARCHIES CROSS-PRODUCTS

Composition may encapsulate a delegate to mimic a parent component. Even in C++, composition may be a preferred design choice despite language support for multiple inheritance. Polymorphic delegates (with interfaces) may offer more options for variant 'parent' behavior. Previous examples of multiple inheritance reused two existing, standalone classes.

What is a streamlined design for combining two class hierarchies? We trace an intuitive example that melds two existing type hierarchies. Assume two defined class hierarchies: Runner (Marathon, Sprinter, Jogger) and Dog (GreyHound, Husky). The goal is to construct a DogWalker type that combines canine and human exercise preferences. Define a multiply inherited child class, DogWalker, matching dog, and runner subtypes so that exercise regimes are compatible.

Blending two class hierarchies is not a trivial endeavor. The multiply inherited child type DogWalker: public Dog, public Runner inherits only the functionality of the basic Dog, not specialized dog breeds, and the basic Runner type, not the variance of a striving Olympian or fitness enthusiast. How would canine or human exercise specialization be replicated inside the multiply inherited child class? What happens if another descendant is define for either Dog or Runner? OCP is violated because the class has to be opened up for specialization and type extensibility.

Using multiple inheritance, there is no easy or maintainable way to combine two existing class hierarchies. How are incompatible combinations, such as GreyHoundMarathon, handled? How many child classes are needed? The definition of five classes, each inheriting from one Runner (Marathon, Sprinter, Jogger) and one Dog (GreyHound, Husky) but excluding GreyHoundMarathon seems like a lot of work. The definition of three different DogWalkers subtypes (distance, fast, steady), each inheriting from Dog alongside one Runner, forces an internal specialization for dog breeds which likely duplicates code. For example, a child class inheriting from Dog and Sprinter would have to internally specialize the Dog component as would a child class inheriting from Dog and Jogger, etc. Moreover, the introduction of a new breed would require altering the DogSprinter and DogJogger and DogMarathon internal code. Likewise, defining two classes each inheriting from Runner alongside one of the two types of Dogs, forces an internal specialization for runners, again duplicating code.

Abandoning multiple inheritance for a type combination of Dog and Runner improves maintainability. Composition with polymorphic delegates supports type extension, that is, the possible continued evolution of the Dog or Runner hierarchies. The addition of a SaintBernard or an UltraMarathon subtype does not derail design. The four design approaches in Example 7.11 are ranked from least to most extensible.

### Example 7.11 C# Composite Class Hierarchies

```
// 1) C++ single inheritance with excessive delegates
//     not maintainable and potentially unstable
//     restriction that only one delegate active
//     at a time
//                 inefficient and not enforceable
// no automatic exclusion => internal check for
// GreyHoundMarathon
// C++ class avoids overhead of heap memory
// (& copy semantics)
class DogWalker1: public Husky
{       Sprinter        s;
        Marathon        m;
        Jogger          j;
        ...
};

class DogWalker2: public GreyHound
{       Sprinter        s;
        Marathon        m;
        Interval        i;
        ...
};

// 2) C++ direct specification of subtype
// without replication
//     supports exclusion - no GreyHoundMarathon,
//     GreyHoundJogger
//     but does not provide extensibility
//         new Runner or Dog subtypes require new
//         class definitions
class DogWalker1: public Marathon { Husky  dog;   ... };
class DogWalker2: public Sprinter { Husky  dog;   ... };
class DogWalker3: public Sprinter { GreyHound
dog;   ... };
class DogWalker4: public Jogger   { Husky  dog;   ... };

// 3) C# single inheritance with polymorphic
// delegate
// delegate construction internal or via
// Dependency Injection
// internally suppress invalid combinations:
// no GreyHoundMarathon
```

```
// ok if Dog hierarchy extended
//     new Runner subtype requires new class
//     definition
// alternatively use Runner delegate and
// Dog subtype as parent
public class DogWalker1: Sprinter { private Dog d; … }
public class DogWalker2: Marathon { private Dog d; … }
public class DogWalker3: Jogger { private Dog d; … }

// 4) C# 'double composition' - two polymorphic
// delegates
// internally suppress invalid combinations:
// no GreyHoundMarathon
// ok if either Runner or Dog hierarchy extended
public class DogWalker
{       private Runner          r;
        private Dog             d;
        …
}
```

The first two designs of Example 7.11 are not extensible. Data member declarations of each specific subtype are redundant since only one subtype should be active for each DogWalker1 etc. object instantiated. To avoid heap memory management, some older C++ class designs worked so directly with specific types. But this approach does not provide the flexibility of pointers or references which can hold the address of any subtype that conforms to the base type (or interface). In modern C++, smart pointers may be used for flexibility alongside the reduction of management responsibility. Due to the limited typing, and inherent redundancy, these designs would not be improved if written in C#. The only advantage that the second design offers is that incompatible combinations are excluded simply by not being defined.

The third design of Example 7.11 uses single inheritance alongside composition of a **polymorphic** delegate. OCP is upheld because the definition of a new subtype from either hierarchy does not affect any existing classes – all remain closed to modification. Invalid type combinations though must be filtered internally. A new Dog delegate subtype will not force change, if Dependency Injection is used for external instantiation of the delegate, and if the new Dog subtype is not incompatible with the defined parent. Similarly, a new Runner subtype requires no modification: a new child class inheriting from the new Runner subtype must be defined. This third design is extensible.

Double Composition may be the most extensible design. Two polymorphic delegates – one Dog and one Runner – support all combinations of Dog and Runner subtypes within a single class definition. A polymorphic delegate is typed to the base class, and may indirectly address an object of any subtype in the class hierarchy. Extensibility is easily satisfied but all incompatible combinations must be screened internally. For example, DogWalker must disallow the combination of delegate r being of type Marathon alongside delegate d being a GreyHound. Since each delegate is replaceable, the designer must consider subtype stability and requirements for replacement.

Any simulation of multiple inheritance must determine what interface the composite class supports – a strict sum of two interfaces is not necessarily required. Overlapping functionality may be combined; some functionality may be melded. Error responses must be defined, or omission contractually noted.

## 7.7 OO DESIGN PRINCIPLE

The **DRY (Don't Repeat Yourself)** principle emphasizes the judicious reuse of code in preference to copy & paste replication which is tedious and error-prone. From the abstract classes in the disassembler example to the simulation of multiple inheritance, designs in this chapter sought to reuse code so as to minimize code replication. In particular, composition designs that use inheritance via polymorphic delegates achieved much flexibility. A virtual call through a polymorphic delegate automatically yields a wide variety of behavior.

## 7.8 SUMMARY

Software design's immediate goal is to meet user expectations by providing required functionality. Other priorities may include performance, security, usability, scalability, maintainability, software product line extension, etc. *A software designer cannot optimize all criteria simultaneously and so must evaluate tradeoffs.* Comparative analyses may be difficult if short-term and long-term priorities conflict. For example: maintainability, via type extensibility as provided by virtual methods, may impact performance; abstract interfaces may increase layering; retention of existing structure, via code reuse, may force wrapping; accommodation of testing and/or environment change may yield Dependency Injection. When different designs are viable, an assessment of costs and benefits may drive selection. Though design decisions may be difficult, competent software designers can simulate missing features and determine when and how to avoid expensive approaches.

Both composition and inheritance reuse code effectively. Pure inheritance, or type extension, rests on interface stability. Support of the is-a relationship suggests type continuity as manipulated through the base class interface. Composition relies less on stability and often wraps unstable interfaces. The relevance of type is a dominant factor in design selection. Data types promote safe and consistent manipulation of memory.

Type combination exemplifies design complexity. Multiple inheritance is vulnerable to two design difficulties: **ambiguity** (overlapping interfaces) and **redundancy** (overlapping type representation). The compiler forces a design resolution to ambiguity. No tool assists with redundancy. Both problems occur when parent type definitions overlap. When two or more parent classes do not overlap in form or function, the parent classes are **orthogonal**. Multiple inheritance designs that are clear, maintainable, and effective arise from orthogonal parents that serve as individual types (not type hierarchies).

## 7.9 DESIGN EXERCISES

The first design challenge is to construct a hierarchy of maps, where each *map* object encapsulates a two-dimensional array of integers, restricting data values, and provides the following functionality:

1. populate an encapsulated 2D array of integers, with special border values

2. conditionally overwrite the value in row x, column y

3. 'freeze' a map location; frozen locations may not be overwritten

4. for a specified row, return the minimum or maximum value to client

5. for a specified column, return the minimum or maximum value to client

Make the base class abstract and define two descendant classes of the basic map type: modMap is-a map that encapsulates a two-digit 'mod' value m used to ensure that all values in the map are evenly divisible by m. *modMap* objects will not freeze any values in a row or column whose indices are evenly divisible by m; and, uniqueMap is-a map that holds no duplicates and will not freeze values.

Since the base class is abstract, inheritance is required to complete any type definition. Type extensions though are uncontrolled and are not guaranteed to be fully utilized through the base class interface. The second and third design problems intend to illustrate difficulties with inheritance designs. Ideally, the reader should consider these problems *only after* constructing a solution to the first problem.

The second design problem is to define subtype `primeMap` that stores only non-primes numbers, limits the number of possible replacements and freezes no more than half the values in the array. The third design problem is to define subtype `thawMap` that has no restrictions on the data values stored and supports the unfreezing of frozen values.

The last design problem appears to be a multiple inheritance design – and that is an option but only in C++. Start by defining two marker types that move across a two-dimensional grid: `inchworm` crawls along a vertical or horizontal line, marking all cells in its path; `leapFrog` jumps from one cell to another along a diagonal, marking only the destination cell. Next, define a `leapWorm` class that crawls along a diagonal, reusing these existing types. Viable design options include: 1) multiple inheritance (C++); 2) single inheritance from one parent alongside composition (the second parent is subordinated to a data member); and 3) double composition with both parents subordinated to data members. Solutions are sketched and discussed in Appendix C.2.


## DESIGN INSIGHTS

*Software*

Compilers do NOT handle ambiguity

Interfaces promote consistency and use of heterogeneous collections

*Software Design*

Poor design cripples the possibilities of extension.

Software complexity often seems unavoidable.

Forced reuse may make design difficult.

Composition provides more flexibility than inheritance

Polymorphic subobjects promote variation

*Documentation*

Identifying optimal designs is difficult => record rationale for choice

Evaluate tradeoffs

Document intent and effect

## CONCEPTUAL QUESTIONS

1. When would suppression of inheritance be appropriate?

2. What are the costs and benefits of multiple inheritance?

3. How can multiple inheritance be simulated?

4. Describe the notion of type subordination and how it affects design choices.

5. When should composition be chosen in lieu of inheritance, and vice versa?

## CHAPTER 8

# Operator Overloading

## CHAPTER OBJECTIVES

- Define operator overloading

- Contrast C++ and C# support

- Consider conceptual framework

- Assess utility of operator overloading

## 8.1 OPERATORS REPRESENT FUNCTIONS

Interfaces should be intuitive. This common observation is easily understood by replacing 'code' with 'interface' in Cory House's wry comment "Code is like humor; if you have to explain it, it's bad". Much work is put into the design of user interfaces to facilitate use. The same should be true of designing a class interface. Overloaded operators, often described as syntactical sugar (sweetens but does not add functionality), can sustain intuitive use of a type.

Overload operators require type resolution of the operands in order to invoke the appropriate implementation – a more efficient and less error-prone process in a statically type language where the compiler resolves type. Casting of an operand may impact function resolution. For example, if x is an integer and y is a real then y may be cast to an integer and added to x via integer addition. In contrast, if x is a real and y is an integer then y may be cast to a real and added to x via real addition. Casting offers

TABLE 8.1  Types of Operators

| Semantic Meaning | Operators | Destructive | Value Returned |
|---|---|---|---|
| Mathematical | `+, -, *, /, %` | No | Temporary |
| Relational/ Comparison | `<, <=, ==, !=, >, >=` | No | Boolean |
| Logical | `&&, \|\|, !` | No | Boolean |
| Increment/ Decrement | `++, --` | Yes | Object |
| Access | `[], ->, *` | No | subObject |
| Function | `()` | Possibly | Varies |
| Stream I/O | `<<, >>` | No/Yes | Stream |
| Assignment | `=, +=, *=, -=, /=` | Yes | Lvalue |

convenience but may causes difficulty when overloading operators for user-defined types, especially with mixed-mode arithmetic.

Operator overloading can increase abstraction and readability: 'x + y' is more readable than 'add(x,y)' or 'x.add(y)'; 'x < y' is more readable than 'isLessThan(x,y)' or 'x.lessThan(y)'. Moreover, overloading an operator permits instantiation of that type in a generic container or algorithm (e.g. '<' for std::sort). Table 8.1 delineates the different types of operators commonly available, identifying return values as well as whether the operations are destructive, that is, alter an operand. Although counterintuitive, for example, '+' is not destructive: neither operand is altered; '+' returns a temporary object as the sum.

C++ permits all but four operators to be overloaded. Java does not support any operator overloading. C# supports limited operator overloading with specific restrictions. C++ implements overloaded operators in two ways: non-static class methods and global methods outside class scope. A C++ binary operator overloaded as a class method is invoked through the left operand (object) and the right operand (object or literal) is passed as a parameter. The this pointer holds the address of the left operand. Non-destructive class methods should be labelled const. A C++ operator overloaded as a global method passes all operands as parameters. C# overloads operators via static class methods, also requiring all operands to be passed as parameters. See Example 8.1.

### Example 8.1  Invocation of Addition

```
x + y;          // operator syntax
add(x,y)        // function call syntax
```

```
x.operator+(y)   // C++ non-static method call syntax
operator+(x,y)  // C# static method call syntax
```

Software design cannot change the compiler: 'x + y' is always parsed as a binary operation; 'a*b + c' is always processed as '(a*b) + c' since multiplication has higher precedence than addition. Neither the parity nor the precedence of an operator may change when overloaded. Compilers are written relative to a language standard which defines the legal use of symbols as operators. A class designer cannot define methods for symbols not used as operators, e.g. '#' is only a preprocessor symbol in C++.

## 8.2  OVERLOADING ADDITION IN C++

Example 8.2 illustrates the overloading of addition, via a class that cycles through an encapsulated sequence, emitting its values one by one upon request. A design goal is to support client use of a sequence object as if it were a primitive type. The cyclicSeq class uses a STL vector to ensure correct memory management. (The STL vector is an effective alternative to raw arrays when size is stable.)

**Example 8.2  C++ cyclicSeq**

```
class cyclicSeq
{       int              index;
        unsigned         size;
        vector<int>      seq;
public:
        cyclicSeq(vector<int> s)
        {    seq = s;
             size = seq.size();
             index = -1;
        }

        int getValue()
        {    index = (index + 1) % size;
             return seq[index];
        }

        // x + y is NOT destructive:
        // x & y not altered
```

```
// return by value returns
// temporary cyclicSeq object
cyclicSeq  operator+
(const cyclicSeq&  b) const
{      cyclicSeq          local(seq);
       int num = size < b.size? size: b.size;
       for (int i=0; i < b.size; i++)
             //#1 equivalent size?
             local.seq[i] += b.seq[i];
       return local;
}
...
};
```

Syntactically, overloading operators is easy: define methods using the keyword operator. Design may not be as simple. Addition of two cyclicSeq objects could be the summation of 'corresponding' values or the union of two sequences. For example, if the first cyclicSeq object holds values 1, 14, 10 and the second holds values 2, 6, -3, the sum of the two could be 3, 20, 7 (value addition) or 1, 14, 10, 2, 6, -3 (union). Example 8.2 takes the former approach and adds corresponding sequence entries.

**Example 8.3  Client Expectations => Potential Difficulties**

```
// client code: getHeapInt(num)
// returns a vector of num ints
cyclicSeq  large(getHeapInt(500));
cyclicSeq  tiny(getHeapInt(5));
cyclicSeq  holder(getHeapInt(100));
                        // #1 capacity inconsistencies
// 5 values of tiny added to large
holder = large + tiny;
// 500 values added to tiny?
holder = tiny + large;


                        // #2 literals are not objects
// value 2 added to values of large
holder = large + 2;
```

```
// large added to literal 2?
holder = 2 + large;


                                 // #3 increment destructive
holder = large + 1;              // ok, large not altered
holder = large++;                // ++ destructive
```

Overloading addition appears trivial, and it can be, but is design consistent? Sample client code in Example 8.3 uses the public interface of cyclicSeq without knowledge of encapsulated details, and illustrates three potential inconsistencies: size; object vs. literal manipulation; nondestructive vs. destructive operations.

Cardinality of any encapsulated sequence is not evident. What if operands are of different (internal) cardinality? If the left operand is smaller than the right, should trailing sequence values be appended, as in statement #1 of Example 8.2? Design should manage size (cardinality, capacity). Example 8.4 modifies cyclicSeq to accommodate size inconsistency: a local cyclicSeq is allocated a size equal to that of the larger operand. Then, element by element, values from the two cyclicSeqs are added together, until every item in the smaller cyclicSeq has been processed. Capacity inconsistencies are problematic in C++ when raw arrays are used, due to the possibility of run-time errors or data corruption if array indices are out of bounds.

A client may wish to add a number to a cyclicSeq object, say increase each element in a sequence by a given value. Example 8.4 includes a second (overloaded) version of addition which adds an increment to each element stored in a cyclicSeq for mixed-mode addition. Note that the first operand is a cyclicSeq; the second operand is an integer value to be added to each element in the left operand. The cyclicSeq operand is not altered: '+'generates a temporary cyclicSeq object that holds the defined sum.

### Example 8.4 Overloaded Overloaded +

```
//updated operator+    -- handle
//different capacities
cyclicSeq   cyclicSeq::operator+(const cyclicSeq&  b)
{    // copy larger vector for local allocation
```

```
    vector<int>      copy = size > b.size?
                              seq: b.seq;
    cyclicSeq        local(copy);

    unsigned    number = size < b.size?
                          size: b.size;
    for (int i=0; i < number; i++)
         local.seq[i] = seq[i] + b.seq[i];
    return local;
}


// overloaded overloaded + : add increment
// to each element seq
cyclicSeq  cyclicSeq::operator+(int increment)
{   vector<int>      copy = seq;
    cyclicSeq        local(copy);
    for (int i=0; i< size; i++)
         local.seq[i] += increment;
    return local;
}


// Addition COMMUTATIVE: a + b == b + a
// GLOBAL function to invert "7 + a"
// invocation via callback
cyclicSeq  operator+(int literal, cyclicSeq b)
{   return     b + literal;              }
```

Logically, 'y + 5' is the same as '5 + y' because addition is commutative. The compiler examines the left operand for type information in order to invoke the appropriate addition function. If left operand 'y' is a primitive, the compiler chooses the addition function for that primitive type (integer, real, etc.). If left operand 'y' is a user-defined type, the compiler looks in the class for a publicly defined operator+(int)method. When '5' is the left operand and 'y' is not a primitive type, there may be a problem. A class method cannot be invoked through a literal! Since 'a + b' represents the C++ function call 'a.operator+(b)', operand a must be an object. Given the two class methods for operator+ shown in Example 8.4, b can be either an object or an integer. However, 'a.operator+(b)' cannot be fulfilled by a class method when a is a literal: '7.operator+(b)' is not a valid invocation of a class method and it cannot be turned into one. Class designers cannot rewrite the compiler.

Global functions are defined outside of class scope and so are not invoked through an object: no this pointer is passed as an implicit parameter. Without the left operand serving as an implicit parameter (the 'this' pointer), both operands for an overloaded binary operator must be passed as parameters. Commutative expectations may be met through a global function: '7 + a' will compile given the last overloaded '+' in Example 8.4. If an operation is commutative, as is addition, the global function simply forwards the call to the class method, reversing the order of the operands as specified by the client. Call inversion removes the need to: 1) write new code; 2) access private data of the operands.

The last noted inconsistency of Example 8.3, non-destructive versus destructive operations, arises from assignment. When a client may use '+' and '=', then support for '+=' is expected; Example 8.5 illustrates a C++ implementation of an overloaded '+='. A C# class designer cannot overload any assignment operator but does not need to do so: '+=' is 'automatically' provided if '+' is overloaded. A C++ class designer must overload every assignment operator that the client may expect to use (=, +=, -=, etc.).

Assignment is implied with support for mixed-mode addition. If a literal number may be added to an object, an increment of '1' is possible. Consequently, support for pre and post increment may be expected. Yet, pre and post increment are destructive operations while addition is a non-destructive operation – a temporary is returned from the addition method. Type definitions appear inconsistent if client expectations exceed functionality provided by the class. A client comfortable using a user-defined type as a primitive type may expect support for all operations in a 'conceptual framework'.

### Example 8.5 Shortcut Assignment

```
// overloaded +=     (shortcut for addition
// AND assignment)
//   a += b is destructive  // same as a = a + b
//   += invoked through object a =>
//       object a altered
cyclicSeq&    cyclicSeq::operator+=(cyclicSeq& b)
{    unsigned  number = size < b.size?
                        size: b.size;

     for (int i=0; i < number; i++)
          seq[i] += b.seq[i];
     return *this;

}
```

'++' is a placeholder for two different actions: pre-increment and post-increment. The compiler uses the placement of '++' before or after a variable to resolve the call to either pre or post increment. Both methods have the same name (operator++) and one implicit parameter (this pointer). To distinguished between pre and post overloaded increment, a C++ class designer defines two different methods, where post-increment has (a dummy) int parameter in order to provide a distinguishing signature. The compiler automatically patched in a dummy int (typically of value 1) when generating the call; implementation code should ignore the dummy int since its value is not guaranteed.

When pre-increment fires, the updated object state is returned to the caller. When post-increment fires, the updated object state is recorded internally but the original state of the object is returned to the caller. In this manner, pre- and post-increment mimic the defined operators for primitives. Example 8.6 displays the standard implementation of pre- and post-increment for an intuitive example (a Clock) as well as client code illustrating the use of pre- and post-increment.

**Example 8.6 Pre- and Post-Increment**

```
// overloading ++: class must
// distinguish between pre & post
// compiler inserts a dummy int
// for post increment call
// two methods:    operator++()
// -- pre-increment
//                 operator++(int)
// -- post-increment
// => 1) same process for decrement;
// 2) client not impacted
class Clock
    {    ...
        // private utility function
            Clock tick();
    public:
        Clock operator++()
            {  return            tick();  }
        // ...
        Clock operator++(int)
        {    Clock      oldState = *this;
            tick();
```

```
                    return        oldState;
            }
};
// ...
// client code
Clock       am(11, 59);      // 11:59 am
int         x = 10, y = 14;

// display1 holds time of 11:59 am
Clock display1 = am++;
                                      // post++: am holds
                                      // time of 12:00 pm
    // display2 holds time of 12:01 pm
    Clock display2 = ++am;

    cout << display1.time()
         << display2.time()   << endl;
    // 11 and 14 output
    cout << ++x  << y++  << endl;
```

Example 8.6 illustrates the general design for overloading the pre-
and post-increment and decrement operators. A private utility function
(e.g. `tick`) increments according to type definition. Pre-increment simply forwards the call to private utility function `tick` which returns a
copy of the 'incremented' object. Post-increment stores a copy of the
original object, calls the private utility function (`tick`) to advance the
state of the object and then returns the copy of the original: the current,
incremented object is not returned because the client chose post (after)
increment.

Comprehensive support for addition may not end here. How many
methods must be defined to thoroughly support addition? When
mixed-mode addition is not supported but assignment is, the C++ class
designer overloads '+' and '+=' to support 'object + object' addition and
'object += object' addition assignment. When assignment and mixed-
mode addition are supported the number of methods overloaded
expands:

object + object object += object

object + literal object += literal

literal + object object++ ++object

If '+' is supported, it may be reasonable to expect support for '-'. Here we go again! With subtraction, one must again consider mixed-mode arithmetic, pre- and post-decrement, and subtraction assignment. Regardless of support for addition or subtraction, the equality operators should be overloaded, and if ordering objects is relevant, the comparison operators as well. Operator overloading can be a complex design problem. Class designers must strive to provide a coherent set of overloaded operators so that clients may manipulate objects consistently.

## 8.3 CLIENT EXPECTATIONS

If 'a = a + b' is supported, 'a+= b' is expected to compile and run correctly. If 'a = a + 1' is a valid statement, then 'a++' and '++a' should be valid. A class should overload a consistent set of operators to provide a conceptual framework. If '==' is overloaded, then '!=' should be overloaded, etc. C# enforces some design expectations. C++ does not. Table 8.2 delineates common operators and their associations.

A class should control the state of instantiated objects. For C++ classes then, destructive operations should be defined as non-static class methods. Non-destructive operators need not be so restricted. An exception is the input stream operator '>>', which must be overloaded as a global function even though, by accepting input, it alters state – see Section 8.5. Table 8.3 delineates recommendations for overloading C++ operators.

## 8.4 OPERATOR OVERLOADING IN C#

C# supports limited operator overloading and defines all overloaded operators as static class methods. All operands must be processed as parameters: there is no `this` pointer representing an operand. Mixed-mode arithmetic presents no design difficulties in C#, unlike C++, since both operands must be passed for a binary operation. The C# class simply

TABLE 8.2　Operators in Conceptual Framework

| Operator | Related Operators | Associated Operations |
|---|---|---|
| + | - | +=, ++, -=, -- |
| * | / | *=, /= |
| << | >> | |
| < | > | <=, >= |
| == | != | |
| && | \|\| | ! |

TABLE 8.3   C++ Operators

| Class Method Only | Global Method Only | Either |
|---|---|---|
| [] | << | +, -, *, /, % |
| () | >> | <, > |
| =, +=, -=, *=, /=, %= | | <=, >= |
| *, -> | | ==, != |
| ++, -- | | \|\|, && |
| ! | | |

defines three overloaded functions, e.g. taking parameters (type, type), (type, int), (int, type). The type amplify in Example 8.7 illustrates C# operator overloading.

**Example 8.7  C# Overloaded Operators: Static Class Methods**

```
public class amplify
{     private uint     scale;
      private bool     on = true;

      public amplify(uint amp = 1)
      { scale = amp % 100;    }

      public bool isOn()
      { return on;                 }
      public bool toggleOn()
      { return  on = !on; }

      public double increase(double x)
      {     if (!on)    return 0.0;
            return     x + (x*scale/100);
      }

      // =>  both operands passed,
      // no implicit this parameter
      // three versions of operator+ to
      // support mixed-mode
      //           += automatically overloaded
      // non-destructive: new amplify
      // object returned
      public static amplify
      operator+(amplify a, amplify b)
      {     return new amplify(a.scale + b.scale); }
```

```
    public static amplify
    operator+(amplify a, int b)
    {    return new amplify(a.scale + (uint)b); }

    public static amplify
    operator+(int b, amplify a)
    {    return a + b;    }

    // one version of ++ --
    // accommodates pre & post
    public static amplify operator++(amplify obj)
    {    amplify local = new amplify(obj.scale);
         obj.scale++;
         return local;
    }

    public static bool
    operator==(amplify a, amplify b)
    { return a.scale == b.scale; }

    public static bool
    operator!=(amplify a, amplify b)
    { return a.scale != b.scale; }

    public static bool
    operator<(amplify a, amplify b)
    { return a.scale < b.scale; }

    public static bool
    operator>(amplify a, amplify b)
    { return a.scale > b.scale; }

    public static bool
    operator<=(amplify a, amplify b)
    { return a.scale <= b.scale; }

    public static bool
    operator>=(amplify a, amplify b)
    { return a.scale >= b.scale; }

}
```

In C#, overloading arithmetic operators automatically supports shortcut assignment. For example, if '+' is overloaded then '+=' is "automatically"

TABLE 8.4   C# Operators

| Pairwise Operators | Arithmetic ( => Assignment) | Logical, Increment And Other Operators |
|---|---|---|
| ==,  != | +,  -,  *,  /,  % | <<, >> |
| <, > | (+=,  -=,  *=,  /=,  %=) | &, \|,!,˜ |
| <= , >= | | ++,-- |

overloaded. Why? By the language standard, C# compilers treat '+=' as two distinct operations: addition followed by assignment. To process '+=', the C# compiler invokes '+', followed by a call to the assignment operator. In contrast, C++ processes shortcut assignment operators as distinct operations. In C++, overloading '+=' is required for design consistency if '+' is overloaded and '=' is supported. Table 8.4 enumerates C# operators that may be overloaded, including parenthetically, the implied short-cut assignment operators.

C# appears to automatically support the equality (and inequality) comparison operators. Actually though, 'x!=y' is just a comparison of addresses since all C# objects are references. For value-based comparison, the '!=' and '==" operators must be overloaded; C# requires that both be overloaded. See Example 8.7. The compiler generates an error if a paired operator is not overloaded. C# enforces conceptual expectations: if 'not equal' is meaningful, then 'equal' must be meaningful, etc. Relational operators in C# must also be overloaded in pairs; an overload of '<' forces an overload of '>', etc.

## 8.5  OPERATORS OVERLOADED ONLY IN C++

Most C++ operators may be overloaded. Direct access, I/O, casting and transparency all are supported. The potential to alter state and the need to access data members, may present obstacles as seen in this section. Contractual design should specify supported operators in the class invariant. The implementation invariant should record the interpretation of an operator's meaning and all details relevant to destructive operators.

### 8.5.1  Indexing Support

Overloading the [] or index operator provides direct data access, similar to an array: 'x = A[i]' extracts an element from array A and assigns that value to the variable x. The overloaded [] operator may include bounds checking, as shown in Example 8.8, a safety feature missing for C++ raw arrays. An exception is thrown if the index value passed is out of range.

The client can manipulate a cyclicSeq object as if it were an array, but without the possibility of data corruption due to over or underflow. Contractual design should specify restrictions, such as the client must handle thrown exceptions for bounds errors.

Why is a reference returned from the overloaded []? An array name is viewed as the address of the first element of the array. Subsequent elements of the array are located using this base address and adding an offset (the product of the index times the element size). With zero-based indexing, A[0] is the address of the first element, A[1] the address of the second element,..., A[i] the address of the i+1$^{st}$ element. When the reference so returned is used in a read-only manner, the cyclicSeq object is not altered, as in statement #1 of Example 8.8. However, the reference returned does yield an addressable memory location which can be overwritten: as shown in statement #2, the cyclicSeq object is altered. Since an external provision of a memory address permits uncontrolled change, the decision to overload operator[] should be made carefully.

**Example 8.8  C++ Index Operator => May Change State**

```
// overloaded operator provide bounds checking
int&        cyclicSeq::operator[](int index)
{    if (index < 0 || index > size)
         throw range_error("out of bounds");
     return seq[index];
}
...
// client code
cyclicSeq       a(getHeapInt(30));

// #1 data retrieved but not altered
cout << a[11] << endl;

// #2 element of sequence altered
a[14] = 77;
```

### 8.5.2 I/O via the Stream Operators

The C++ stream operators '<<' and '>>' may be overloaded to facilitate use of output and input streams. Again, designers must address the quandary of supporting a destructive operator. Access to private data members must be granted because '<<' and '>>' cannot be overloaded as class methods:

the left operand is an IO stream. Since the C++ utility classes cannot be opened up to add overloaded methods for user-defined types, overloading the stream operators requires the definition of global functions. However, unlike mixed-mode arithmetic, the call cannot be simply inverted.

How do global functions operator>>(cin,object) and operator<<(cout, object) access an object's private data members? Class designers should not make private data members public just to accommodate overloaded stream operators. Public data members violate encapsulation and information hiding, making every object vulnerable to uncontrolled change. The answer is friends. The C++ friend construct permits controlled external access to private data and functionality. Using the reserved word friend, a class designer selectively denotes which external functions, and/or classes, are privileged with private access. Declaring an external class a friend is bolder than identifying a single function: if class TypeX is declared friend of class TypeY then all methods in class TypeX have access to all private methods and data members of class TypeY.

**Example 8.9  Overloading C++ Stream Operators: Friends**

```
class cyclicSeq
{     ...
      unsigned         size;
      vector<int>      seq;
      friend ostream& operator<<(ostream&,
                       const cyclicSeq&);
public:
      ...
};


...
// function must be 'friend' to access
// private 'size' and 'seq'
ostream& operator<<(ostream& out,
         const cyclicSeq& c)
{     for (int i=0; i < c.size; i++)
           out << c.seq[i];
      out << endl;
      return out;
}
...
```

```
// cin >>object;     cout << object;
// more abstract client code than
// object.input();   object.display()
```

Friend declarations may be placed anywhere in a C++ class header file – either the public or private section. Despite the compiler's indifference, placement should be consistent for code readability, and documentation should note external support for overloaded operator(s). Example 8.9 defines a global function for overloading the output operator. To access private object data, the global function must be declared a friend.

The friend construct violates encapsulation, selectively exposes a class and increasing coupling. Yet, restrictions on the friend construct diminish its violation of encapsulation. Friendship is not transferable or assumable. Friendship is not transitive: if A is friend of B and B is friend of C, A is not a friend of C unless C explicitly declares A as a friend. Friendship is not be inherited: if A is friend of Parent, A is not friend of Child, unless Child also declares A as a friend. Friendship is not symmetric: if A is a friend of B, B is not friend of A, unless class A also declares B as a friend. The C++ friend construct allows the class designer to control access through explicit labelling.

The friend construct is controversial. For software maintainability, the class designer should document all friendships, typically in the implementation invariant. The friend construct is necessary to support mixed-mode operations and the stream operators.

### 8.5.3 Type Conversion

Data types promote safe and consistent manipulation of memory. Chapters 6 and 7 examined *type reclamation* from base class pointers and references when the base class interface was insufficient. Both C++ and C# provide operators to do so. static _ cast and as efficiently check type at compile-time but are prone to error if the extracted type does not match the specified cast at run-time. dynamic _ cast and is check type at run-time in order to avoid such errors. C++ permits the overloading of '()' in order to support type casting (as distinguished from type reclamation).

Casting is the action of converting the value of one type to the equivalent value of another type. Most everyone is familiar with casting an integer to real, as in 3 becomes 3.0. Casting a real to an integer is often accomplished via truncation. Implicit casting or type coercion is an action undertaken automatically by the compiler. Explicit casting or type conversion is by directive, as shown in Example 8.10. While type conversion is supported for primitives, C++ class designers may support comparable type conversion by overloading the '()' operator. Example 8.10 also shows the overloaded operator for converting a cyclicSeq object to an int.

**Example 8.10  Converting cyclicSeq Object to Int Value**

```
// overloading type conversion operators
int         i;
float       f;
f = i;          // implicit conversion
f = (float) i;  // explicit conversion, C-style
// explicit conversion, functional style
f = float (i);

//    cyclicSeq object => int
//         method returns type (converted value)
//         non-destructive:
//         object state not changed
// overload type conversion:  operator othertype();
//         operator  int();

// conversion operator
cyclicSeq::operator int()
{    int sum = seq[0];
     for (int i=1; i < c.size; i++)
          sum += seq[i];
     return sum
}
```

## 8.5.4 Transparent Access

Overloading C++ operators for memory management is laborious, and relevant only to applications that must closely manage memory alongside tight performance criteria [Loshin99]. C++ class designers may overload the new and delete operators for efficiency. The general idea is to manage a cache internal to a class so that all but the first call to new and the last call to delete may be intercepted and handled locally. Overhead is reduced by circumventing most of the run-time calls to the heap allocator and deallocator. Yet, management of a local, internalized cache is not trivial, and is not recommended. Ownership of heap memory is shared and transferred via smart pointers and move semantics in modern C++, thereby safely promoting efficiency and reducing the need to overload the new and `delete` operators.

Prior to C++11, class designers had to define their own smart pointers. When doing so, the access operators `operator->` and `operator*` were overloaded, as shown in Example 8.11, for transparency. With the access operators overloaded, the client could manipulate the wrapped pointer as if it were raw. The generic definition of smart pointers essentially overloads `operator->` and `operator*` for transparency.

**Example 8.11  Transparent Access**

```
class SmartPtr
{           Type*          ptr;
      public:
               SmartPtr(Type*& p): ptr(p) { p = 0; }
               ~SmartPtr()                { delete  ptr; }

               ... // copying etc

               Type* operator->()    {      return  ptr; }
               Type& operator*()     {      return *ptr; }
};
```

## 8.6 OO DESIGN PRINCIPLE

Operator overloading elevates the intuitive manipulation of a defined type, allowing clients to manipulate objects as primitives. Design emphasis rests on the interface, supporting the **PINI (Program to Interface Not Implementation)** principle. Clients remain unaware of encapsulated details through judicious design and, in C++, careful use of the friend construct.

## 8.7 SUMMARY

From a design perspective, overloading operators establishes a class interface with 'built-in' support for primitive operations. The specification of a complete and consistent conceptual framework for such overloading is not trivial. Operator overloading can effectively increase abstraction and code readability, promoting maintainability. However, operator overloading may be viewed as just syntactical sugar – it just makes code look prettier. Not surprisingly then, different languages provide different levels of support for operator overloading.

C++ broadly supports operator overloading, permitting the overload of all operators except four:

?: (ternary conditional)

:: (scope resolution)

. (member access)

.* (member access through pointer)

Design difficulties associated with operator overloading in C++ include inconsistent management of programmer expectations as well as violation of encapsulation through the use of the **friend** construct. All C++ operators are inherited EXCEPT the assignment operator. Note though that if a child class extends the parent class by adding data that is relevant to any parent overloaded operator(s), then the child class should overload the same operator(s) to accommodate the manipulation of the child data members.

Java does not support any operator overloading. C# partially supports operator overloading. In C#, one can overload the binary arithmetic operators, the shortcut increment and decrement, some logical operators as well as the comparison operators. However, all operators overloaded in C# are static methods.

C++ provides only a bitwise copy for the assignment operator, by default. C# does not have explicitly overload short-cut assignment operators because '+=' is a combination of the two operators '+' and '='. Mixed-mode arithmetic complicates design, as does assignment, short-cut assignment and internal capacity constraints. Once one operator is overloaded, the client may expect similar or associated operators to also be supported.

## 8.8 DESIGN EXERCISE

The primary motive of overloading operators is to increase abstraction and facilitate use of a type definition. Consequently, design should focus on building a consistent framework. This chapter's design exercise is transform the C# amplify class, defined in Example 8.7, to a C++ implementation. Since C++ supports more extensible operator overloading, consider what other operators should be overloaded. A design solution is sketched in Appendix C3.

## DESIGN INSIGHTS

*Software*

C# and C++ process assignment differently

Comparison of C# objects are address-based

=> overloaded C# operators may provide value-based comparison

Smart pointers overload `operator->` and `operator*` for transparency

*Software Design*

Overloaded operators may increase abstraction and readability

Clients expect consistency

=> if '==' is overloaded, then '!=' should be overloaded, etc.

Overloaded operators support type use in generic algorithms (and containers)

## CONCEPTUAL QUESTIONS

1. When is operator overloading appropriate?

2. Why would it be unlikely for a class design to overload just one operator?

3. How many operators should a class overload?

4. What are the benefits and vulnerabilities of the C++ `friend` construct?

5. Describe the key differences between C# and C++ operator overloading.

# Appendix A: The Pointer Construct

## A.1 POINTER DEFINITION

High-level languages rest on abstraction. Design usually proceeds without much attention to system hardware or memory allocation and deallocation. Software development is thus faster and yields more maintainable and portable code. C provided the pointer construct to retain the ability to manipulate memory addresses. C++ supports the pointer construct to be backwards compatible with C and to retain its emphasis on performance.

A pointer holds a memory address. Example A.1 shows sample pointer declarations, initializations, and use. Figure A.1 sketches corresponding, sample memory assignments for variables data and myPtr. Initially, both variables are declared but not defined. That is, no value is assigned to either variable, as represented by question marks in the diagram. Actually, there are always values in memory. If variables are not initialized, residual bit strings lingering in memory may be erroneously interpreted as valid data. C# and Java zero initialize data declarations. C++ does not. C++ design guidelines recommend initialization but compilers do not enforce convention. Uninitialized pointers are particularly dangerous since a wildcard bit string could be interpreted as a valid memory address. Statements #4 and #5 show initialization after declaration: data holds value 100; and myPtr holds the address of data (B500).

Operators '*' and '&' appear throughout the code of Example A.1. '*' is the indirection operator, it is used to define pointers and to extract data values from the memory addressed by a pointer. '*' does not distribute: statement #3 declares two variables, int pointer iPtr and int x. '*' may be placed immediately after a typename (statement #1) or may precede the variable name (statement #2) in a declaration. The latter style is often

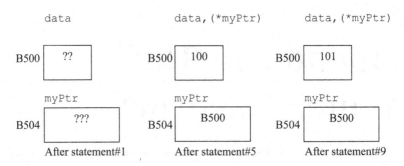

FIGURE A.1    Memory Sketch for Example A.1

preferred because of the lack of distribution but neither style is enforced. For consistency, developers should adhere to one convention.

Statement #1, "`int* myPtr;`" declares a pointer variable that is "typed" to hold the address of an integer: it may not hold the address of any other type. Typed pointers promote consistency. `int` pointers may hold only addresses of `int` variables; `float` pointers may hold only addresses of `float` variables, etc. Type incompatibilities trigger compilation errors. Any statement assigning the address of an `int` variable to a `float` pointer will not compile, etc. This restriction is relaxed with inheritance: a base class pointer may hold the address of a derived class object, as seen in Chapters 5–8.

Untyped (`void`) pointers were broadly used in C code. The address of any type may be held in a `void` pointer. Statement #2 declares a `void` pointer `typelessPtr` that, starting at statement #6, is assigned three different addresses: the address of `int` variable `data` (B500), then the address of a `float` variable, and, finally, the address of pointer variable `myPtr` (B504). `void` pointers provide flexibility but mask type so the client must extract type information when needed. C++ discourages but still supports the use of `void` pointers.

'&' is called the address of operator, and may be used to extract the address of a variable. Statement #4 "`myPtr = &data;`" places the address of variable `data` (B500) in pointer variable `myPtr`. '&' is also used for call by reference. Function header "`void passByRef(int& formal)`" specifies that the formal parameter `formal` is not allocated its own space; upon function invocation, `formal` is aliased to the actual argument. Function call "`passByRef(x)`" thus causes `formal` to reference the same memory location as x. Call by reference is efficient (no extra space or copying overhead) but insecure since modifying the `formal` parameter alters the actual argument as well. Unless declared `const`, pointer values may change, just like any other type of variable. A variable's address though represents assigned data storage which cannot change if determined by the compiler. As shown in Figure A.1, the values held in myPtr and data may change, but their assigned memory locations do not.

## A.2 DEREFERENCING POINTERS

Placing '*' in front of a pointer variable 'dereferences' the pointer: it tells the compiler to manipulate the value located at the memory address held in the pointer. Contrast the output of data values versus addresses of variables in statements #7 and 8. *myPtr and data are aliases, referring to the same memory, because myPtr holds the address of data. A change to either *myPtr or data is equivalent to a change to both. Altering *myPtr modifies data (and vice versa) as long as myPtr holds the address of data. The integer value *myPtr (100) output in the cout statement #7 is the same as that output in statement #8.

**Example A.1 Declare, Initialize, and Use Pointers**

```
// C++ variable declarations
// => all allocated on stack
int     data;
float   realV;
int*    myPtr;          // int pointer              #1
void    *typelessPtr;   // void pointer             #2
int*    iPtr, x;        // int pointer and int      #3

// data initialized with value 100                  #4
data = 100;
myPtr = &data; // myPtr initialized
               // with address of data              #5

typelessPtr = &data;     // void pointers           #6
typelessPtr = &realV;    // may hold address of
typelessPtr = &myPtr;    //any type
//cout value stored in memory followed
//by memory address
//    100 B500 is output                            #7
cout << data << &data << endl;
//    100 B500 is output                            #8
cout << *myPtr << myPtr << endl;
```

```
(*myPtr)++;                      // 101 B500 output          #9
cout << *myPtr << myPtr << endl;
iPtr = myPtr;

*myPtr++ = 42;                                               #11
*myPtr = 4242;            // what is output below?           #12
cout << *iPtr << iPtr << endl;                               #13
cout << *myPtr << myPtr << endl;                             #14
```

Both *myPtr++ and (*myPtr)++ are valid statements but their effects are different. The post-increment operator ++ has higher precedence than the dereferencing operator *. Parentheses may be used to circumvent precedence constraints, as shown in statement #9. (*myPtr)++ forces dereferencing to occur before the post-increment operator, yielding, as desired, an incremented integer value. When (*myPtr) is incremented, the value altered is the data in memory location B500 because B500 is the value (address) held in myPtr. Hence, data becomes 101. The value of the pointer variable myPtr is not altered; the value held in the memory that it addresses is.

Without parentheses to circumvent precedence, *myPtr++ in statement #11 invokes the post-increment operator first, incrementing myPtr. Yes, the value of the pointer variable, an address, is incremented (by '4', assuming pointers allocated 4 bytes – verified, C++17 via CLion). *myPtr++ first increments the value of myPtr, B500, to B504, followed by assignment of value 42 to the integer held at address B500 (the address increment is post not pre). Statement #12 is particularly dangerous because the value 4242 is assigned to a pointer (possibly yielding a run-time error, or data corruption). It is unclear what data could be erroneously altered at memory location 4242. Overwriting addresses held in references and pointers is a tactic of malicious software.

Syntactically, it does not matter what memory is addressed by a pointer. However, at run-time, memory accessed must be within the range of addresses allocated to the running program. If a pointer variable contains an invalid address, that is, a value outside the range of addresses assigned to the executing program, dereferencing that pointer triggers a run-time error. The dereferencing of myPtr discussed in the previous paragraph could yield a run-time error if 4242 was not a valid program address. Abstractly, it is not easy to distinguish invalid from valid addresses. A pointer variable should be initialized to the address of a variable, or to zero to indicate that it currently does not contain a valid address. C and legacy

C++ often initialized pointer variables to zero, or used a defined constant, such as NULL. Modern C++ uses nullptr; C# uses null.

## A.3 INAPPROPRIATE USE OF POINTERS

Example A.2 illustrates illegal and unwise manipulations of pointers. Figure A.2 shows the corresponding memory allocation. A pointer may not hold the address of a constant value or a literal value. Why? There is no program memory associated with constants or literals. The strike-through statements in Example A.2 indicate compilation errors. More problematic are statements that compile but have unknown effect because the pointer variable may or may not contain a valid address (statement #3).

**Example A.2  Danger of the Uninitialized**

```
// C++ code: missing and illegal initialization
float*      fPtr3; // uninitialized pointer variable
int*  iPtr3 = nullptr; // pointer initialized to zero
const float   pi = 3.14159;
*iPtr3 = 15;   // do not deference null pointer!   #1
fPtr3 = &pi; // pointer cannot hold address of constant
// pointer cannot hold address of literal
iPtr3 = &1000;

if (iPtr3) *iPtr3 = 100; // Safety check. Effect? #2

*fPtr3 = 99.99;        // Effect?               #3
```

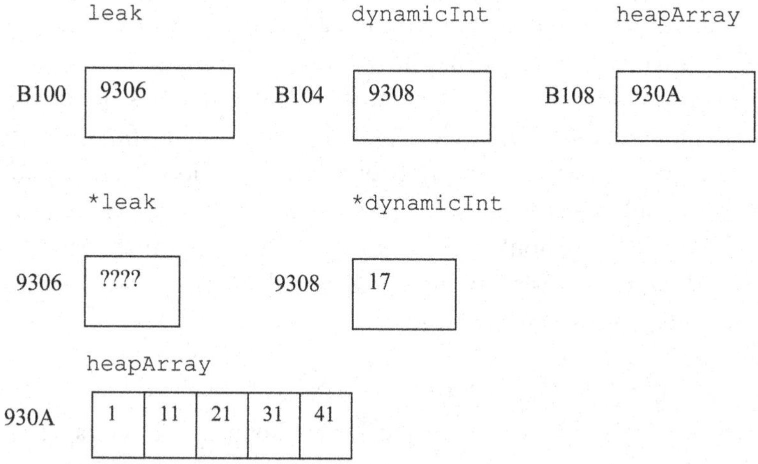

FIGURE A.2    Stack Pointers and Heap Data

Following convention, iPtr3 was initialized to nullptr (points to nothing), indicating that it does not contain a valid memory address. Dereferencing a null pointer, as in statement #1, yields a run-time exception. Why? Memory location 0 is in the operating system domain and is inaccessible to user programs. Exceptions may be used to preserve integrity. It is preferable to generate a run-time exception (which can be caught by exception handling code) from a null pointer than to permit data corruption. Alternatively, a simple safety check may verify that a pointer is not null or zero, as in statement #2. When the conditional fails, dereferencing a null pointer is prevented. Thus, 100 is not assigned to the memory (location 0) 'pointed to' by iPtr3.

What is troubling about statement #3 in Example A.2? fPtr3 was not initialized. Failure to initialize does NOT yield variables with no value, just uncertainty. No one knows the value stored in an uninitialized variable: whatever bit string resides in the memory associated with the variable may be interpreted as valid data. Thus, whatever bit string resides in the memory associated with the pointer fPtr3 will be interpreted as an address of a float. If the residual bit string yields an address outside the range of valid program addresses, a run-time error occurs. Otherwise, the addressed memory is overwritten by value 99.99. Data corruption! Although erroneous memory overwrites do not always affect running software, data corruption can lead to failure. Such errors are often hard to trace because they may occur far from the source. Design guidelines explicitly recommend that programmers initialize pointer variables either to the address of an appropriate variable or to zero (or nullptr) when pointers point to nothing. Modern IDE (integrated development environment) flag the use of pointers before initialization but legacy code may not have been so vetted.

Pointers support the sharing of data. Multiple handles (pointers) can hold the same address value. Pointers may lead to data corruption if the data value in memory is changed through an alias without the knowledge (or permission) of other aliases. The same is true for references. Hence, software developers should track who 'owns' allocated memory. Chapter 2 covers design responsibility for memory in detail. Here, we briefly summarize managing memory allocated at run-time.

## A.4 TRANSIENT VERSUS PERSISTENT MEMORY

Every function has its own memory for local data: parameters passed by value and variables declared in scope are stored in a stack frame associated with the function. The compiler efficiently supports local memory but access to such data is restricted. Upon function entry, a stack frame is pushed onto the run-time stack; upon exit, this stack frame is popped off. Thus, local data 'disappears' when function scope is exited and the memory previously occupied by its stack frame is reused. Local data is inaccessible after a function terminates but copies of data may be explicitly returned from a function call.

What if persistent data is needed? Global data may undermine integrity and confound linkage. Instead, programmers may store data on the heap; retaining handles to such memory allows data to be accessed across multiple scopes. C++ programmers access heap data via pointers but, unlike Java and C# programmers, retain responsibility for the explicit release of such data.

Two operators are used for managing heap memory: new is invoked to satisfy a run-time request for memory; delete is invoked for the run-time release of memory. Both operators depend on pointers. new returns the address of allocated memory; a pointer should hold that address so that delete may be subsequently called to release the heap memory. When heap memory is allocated via new but the address returned is not retained, the programmer loses access to the heap memory. For example, the declaration of variable a via "queryCount a = *(new query-Count(12))" is assigned the value of a dereferenced 'anonymous' address – a heap object (initialized via its constructor). How can the heap memory be deallocated after this statement? It cannot because there is no pointer variable retaining its address. Chapter 3 briefly discusses smart pointers which provide a C++11 solution to such memory leaks.

Example A.3 shows sample allocations and deallocations of heap memory using new and delete. Three pointer variables (residing in a stack frame) hold addresses of heap allocated memory. The heap integer whose address is held in pointer leak in Example A.3 is not deallocated. Hence, the memory so assigned may not be reused by any subsequent new request during program execution. Figure A.2 sketches memory assignments for Example A.3, using addresses that start with 'B' to indicate stack memory and addresses that start with '9' to indicate heap memory. Stack memory, handled via stack frames that are pushed (popped) upon function entry (exit), is efficiently and securely processed: it does not leak. Heap memory is more difficult to manage.

To prevent memory leaks, every new should be matched with a delete, that is, every memory acquisition should be released. In confined scope, it is easy to match every new with a delete (though throwns exceptions

may interfere with this local pairing). In a broader context, it is more difficult to match every new with a delete. Since a key motive for using heap
memory is data persistence, addresses (pointer values) are often passed in
and out of functions. Hence, the call to delete may occur far from the
call to new, making such matching challenging. Chapters 2 and 3 explore
design responsibilities for memory in more detail.

### Example A.3 Allocation and Deallocation of Heap Data

```
int*    leak = new int; //single int allocated on heap
int*    dynamicInt = new int;
                        //array of 5 ints allocated
int*    heapArray = new int[5];

*dynamicInt = 17;       //single int assigned value 17

for (int k = 0; k < 5; k++)       // array initialized
        heapArray[k] = k*10 + 1;

delete          dynamicInt;       // memory release
delete[]        heapArray;        // array released

// leak NOT deallocated => memory leak
```

## A.5 REFERENCES

Like a pointer, a reference is an address holder. Example A.4 illustrates
the declaration, initialization and use of C++ references. Figure A.3 shows
corresponding memory allocations. A reference is manipulated in the
same syntactical manner as the variable with which it is aliased. Variables
z, alias, *iPtr and in Example A.4 all refer to the same data because all
three variable reference the same memory location. Thus, the three output
statements in this example will print out three copies of the same value:
first, the value of the residual bit string in the shared memory location;
then the value 100; and finally, the value 1.

### Example A.4 References

```
// C++ code: aliases (two or more handles
// point to same memory)
int             z;
                // reference to int variable declared
```

```
int&        alias = z;
            // compile-time error: no alias specified
int*  iPtr = &z; // int pointer declared and defined
// #1 Initial stack allocation complete here

z++;
alias++;
cout << z << alias << *iPtr << endl;
// #2 post increment of (uninitialized)
// variables complete here

z = 100;
cout << z << alias << *iPtr << endl;

alias = 1;
cout << z << alias << *iPtr << endl;
// #3 assignment to alias complete here
```

Like pointers, references may yield corrupted data if not handled appropriately. Consider return by reference, illustrated in Example A.5. returnByRef() returns the address of a variable. When a function terminates, its stack frame is popped off the run-time stack and all local variables 'disappear'. A returned address that references a local variable then is invalid. The caller might think the address is valid and, hence, may unwittingly access or modify data values located at that address. Modification may lead to data corruption. When returning a reference, programmers must ensure that the address returned is currently valid, and will remain

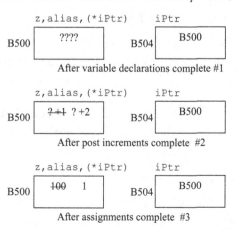

FIGURE A.3   Memory allocated for one int and one pointer

valid after function scope is exited. Local variables go out of scope when a function is exited and so should not be returned by reference. Again, modern IDEs may flag such errors and issue a warning.

**Example A.5  Do NOT return local variables by reference!**

```
int&    returnByRef()
{       int     local;   // stored in stack frame

        cin >> local;
        if (local % 2)     local++;

        return local;
}
// client code
int& alias = returnByRef(); //access memory in
                           //popped stack frame
```

Addresses of persistent variables may be returned from a function. Example A.6 shows the definition of a persistent int on the heap. Passing out the address of heap data is valid because heap memory remains allocated until explicitly deallocated via a call to delete. The caller assumes responsibility then for the heap memory transferred out of the function. Example A.6 is overkill: it is much simpler and cheaper to directly copy an integer value. Preserving heap memory allocation, and transferring ownership via pointers, though yields significant savings when data so stored is large because the overhead of allocation and initialization is reduced.

**Example A.6  Heap Objects Persist**

```
int*    returnPtr()
{       int*    localPtr;   // pointer stored
                            // in stack frame

        int     stackObj;   // int stored in stack
                            // frame

        localPtr = new int;  // int stored on heap
        return localPtr;
}

// client code
int* persist = returnPtr(); //object still on heap
```

Aliases become invalid when scope exceeds the lifetime of allocation. Clients may not recognize such discontinuity. Example A.5 demonstrated this problem via poorly designed return by reference: guidelines mandate that one should not return local variables by reference. If a caller holds an address to memory in a stack frame, the caller is left with a handle to memory that will be reassigned, providing an unchecked opportunity for data corruption. In contrast, Example A.6 returns the address of a persistent, heap allocated object, ensuring valid access.

## A.6 THE this POINTER

Object-oriented design encapsulates data alongside privileged access to defined functionality via the class construct. Functions defined within classes, often called methods, lie within in class scope. If a test() function is defined in the queryCount class, its full name would be queryCount::test(). If a test() function is defined in a track-Mean class, its full name would be trackMean::test(), etc.

Example A.7 defines a C++ class that tracks the number of queries made of an object. For clarity, most other functionality has been omitted. Object instantiation for three variables is also presented, and Figure A.4 shows sample memory allocation for these objects. How does the queryCount::test() method invoked in statement #2 of Example A.6 know that it should update the hit count of object a? Likewise, how does the queryCount::test() method invoked in statement #3 know that it should update the hit count of object b?

**Example A.7  Class Definition => need this pointer**

```
class queryCount
{       int     count;
        int     min;
    public:
        queryCount(int x = 1)
        {       count = 0;
                min = x;
        }
        void    test(int min)
                        // #1 this disambiguates min
        {   if (min > this->min)
                count++;   //same as (min > (*this).min)
        }
};
```

```
// ... client code
queryCount      a, b(2), c(77);
a.test(5);                // #2 equivalent to
                          // queryCount::test(B100,5)

for (int k = 0; k < 10; k++)
        b.test(k);        // #3 equivalent to
                          // queryCount::test(B108,k)

for (int k = 0; k < 100; k++)
        c.test(k);        // #4 equivalent to
                          // queryCount::test(B110,k)
```

The compiler translates the method invocation a.test() to queryCount::test(&a). That is, the compiler passes in an implicit parameter – the address of object a, ensuring that a.count is incremented. Similarly, the compiler translates the invocation b.test() to queryCount::test(&b)– the address of the object b is passed. When count is now incremented, it is b.count++. The implicit parameter is called the this pointer. A standard definition of the this pointer is the address of the object through which a class method is invoked.

The this pointer is defined for each object instantiated, whether code is written in C++, C# or Java. The this pointer resolves access to data members and member functions defined in a class but referenced through distinct objects. The this pointer may be used to disambiguate references. In a class method, formal parameters often carry the same name as the associated data field. One cannot use the same name in one scope to reference different memory locations. The this pointer clarifies that the field referenced is a data member of the object through which the method was invoked, as shown in statement #1 of Example A.7. As always, a pointer must be dereferenced before accessing the data it addresses. An equivalent, and more readable, syntax is achieved by using the -> operator. "(*this). min" is equivalent to "this->min".

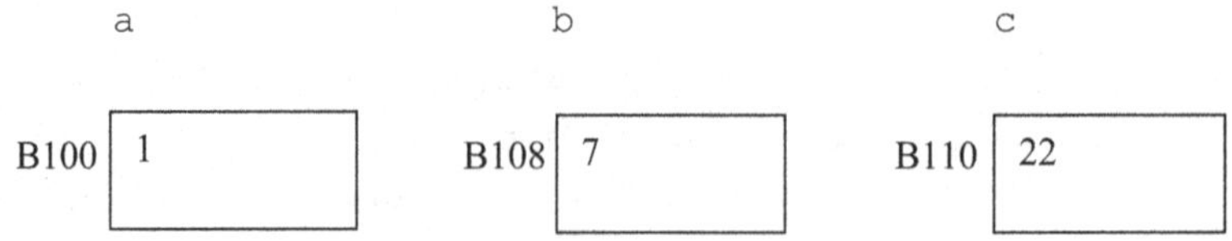

a                              b                              c

B100 | 1                  B108 | 7                  B110 | 22

FIGURE A.4   Three queryCount objects

## A.7 ARRAYS

Fixed-sized array allocation is easily managed in a stack frame. If variable-sized is needed, often an STL container is used. Here, we examine the use of heap memory, calling new (and then delete) for variable-sized arrays, in order to reinforce understanding of dynamic allocation. Example A.8 shows several array declarations. The first is a standard definition. The second is also a fixed-sized allocation but the data type of the array is a pointer. When the compiler allocates an array of objects, it also generates code to invoke the no-argument constructor to initialize each element (object) in the array. Since noObjArray provides only a constructor that takes an integer, the compiler is unable to resolve the array declaration because there is no no-argument constructor to call. In C++, to support array declarations, every class must define a no-argument constructor or provide no constructors (so that compiler provides a default, no-argument constructor). Unable to declare an array of noObjArray objects, a programmer can instead use an array of pointers, as shown in Example A.8.

**Example A.8  C++ Array of Pointers**

```
const int size = 100;
int     fixedSize[size];          // #1   standard array
...

class noObjArray
{       int     secret;
    public:
        noObjArray(int x)   { secret = x; }
};


noObjArray*   db[100];  // #2   array of 100 pointers

// #3 array initialized with addresses of
// heap-allocated object
for (int j = 0; j < 100; j++)
        db[j] = new noObjArray(j);


noObjArray**            ptrPtr; // #4   pointer to pointer
// #5 pointer holds array address
ptrPtr = new noObjArray*[100];
// #6 array initialized with addresses of
// heap-allocated object
for (int j = 0; j < 100; j++)
        ptrPtr[j] = new noObjArray(j);
```

```
                            // #3B deallocate heap objects
    for (int j = 0; j < 100; j++)
          delete db[j];

                            // #6B deallocate heap objects
    for (int j = 0; j < 100; j++)
          delete ptrPtr[j];

    delete ptrPtr;          // #4B deallocate heap objects
```

When using heap objects, every allocation must be matched with a deallocation. In Example A.9, statement #2 allocates an array of 100 pointers; the pointers are initialized in statement #3 to each hold the address of a heap object. Statement #3B provides the corresponding deallocation of each individual heap object in another for loop. Array db, assigned memory in the stack frame, is automatically released when scope is exited. Statement #4 is the declaration of a 'pointer to a pointer'; ptrPtr holds the address of another pointer; in this case, the address of an array of pointers. Statement #5 is the declaration of a heap array of 100 pointers. Statement #6, like statement #3, initializes each pointer in ptrPtr[] to hold the address of a heap object. Statement #6B, like statement #3B, provides the corresponding deallocation of each individual heap object. Statement #4 deallocates the heap array of pointers. Note that statement #5B must run before statement #4 so that the heap objects are deallocated before the pointers that hold their addresses. Figure A.5 contrasts the allocation of an array of objects an array of pointers.

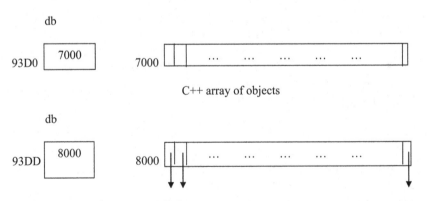

FIGURE A.5    C++ array of pointers initialized to hold addresses of heap objects

A stack-allocated array may be used when a class has a no-argument constructor but the client wishes to initialize each object in an array using a different constructor. Since the compiler always patches in calls to the no-argument constructor for each array element, the client must overwrite the default initialization, as shown in Example A.9. This approach is easier and cheaper than using an array of pointers because there is no run-time overhead for allocating and deallocating memory.

**Example A.9  C++ Array Allocation**

```
// #1   array of objects allocated on stack
//      default constructor implicitly invoked
//      for each object
// want to overwrite NO-ARGUMENT constructor
// default of 1
queryCount           db[100];

for (int j = 0; j < 100; j++)
{       queryCount local(j); // non-default constructor
        db[j] = local;       // #2
}
```

## A.8 SUMMARY

C and C++ support the indirect addressing of memory via the pointer construct. C# and Java do not. Another key difference between C++ and C#/Java is the allocation and deallocation of objects. C# and Java allocate all objects on the heap, implicitly using references. C++ allocates objects on the stack by default and on the heap upon request. C++ heap memory is directly accessible via the pointer construct, yielding more control but also more vulnerability.

The responsibility to deallocate heap memory in C++ is significant. 'Match every new to a `delete`' is a simplistic mantra that is difficult to follow in the midst of function calls and aliases. Aliases, transfer of ownership, parameter passing, etc., make such a naive design guideline difficult to follow, especially without adequate documentation AND in the context of modern, large-scale software. Aliases to heap memory should be stable. Nonetheless, in C++, one must track ownership of heap memory so memory leaks and data corruption do not occur.

# Appendix B: Design Exercises

**Design Choices and OOD Principles for Chapters:**

1. Contractual class design    *Single Responsibility Principle*

2. Ownership (Memory)         *Responsibility Driven Design*

3. Data Integrity (Copying)   *Principle of Least Knowledge*

4. Composition              *Dependency Inversion Principle*

5. Inheritance                *Liskov Substituability Principle*

## B.1 CONTRACTUAL CLASS DESIGN

Chapter 1 examined contractual class design, summarizing goals for type definitions as well as client responsibilities. Class design defines key functionality, accessibility, and state transitions that preserve internal control. Consistency of private data and functionality are not client responsibilities. The class construct *publishes functionality but encapsulates data*, decoupling the client from implementation details and ensuring that objects remain in a consistent, legal state. Standard class functionality includes constructors, accessors, mutators, private utility method, public interface methods, and, possibly for C++, destructors.

Chapter 1's design exercise was to define a class inRange to track the number of integers that fall within a specified range. For example, given a range of 100 to 900, rangeObj.query(117) yields true, rangeObj.query(11) yields false, rangeObj.query(717) yields true. After these three queries, the count of integers that fell within range would be two. As noted at the end of Chapter 1, this problem description is inadequate – many details are missing. Is the range inclusive of its boundaries, [100,900], or exclusive, (100, 900)? One class design can support both by using a Boolean, set in the constructor and thereafter not modifiable. Another Boolean is needed to control on/off state.

Data members needed for inRange objects include: boundary values; a flag (Boolean) to indicate whether the range includes the end values, for example [100, 900] includes 100 and 900 while (100, 900) does not; and another Boolean to indicate whether an object is on or off. Methods defined for inRange include: a constructor to accept boundary values; accessors to retrieve state (on/off) and the query count; a mutator to change on/off state; reset capability to restore the object to its initial state but not modify boundary values; and the main functionality to determine if a given number lies within the range or not.

Tracking the number of values queried that fall within a given range does not require storing data. Example B.1.1 gives a C# solution; there would be no interesting differences in a comparable C++ solution. The range is stable: the upper and lower bounds do not change after object instantiation. C# supports data stability through the keyword readonly – a value cannot change after first assignment. C# also zero initializes data members by default.

### Example B.1.1 C# Class with On/Off State

```
public class inRange
{       private bool          on = true;
        private uint          count;
        private readonly bool inclusive;
        private readonly int  lowBound;
        private readonly int  upBound;

        public inRange(int  x, int y,
        bool border = true)
        {       inclusive = border;
                lowBound = x < y ?  x: y;
                upBound =  x < y ?  y: x;
        }
        public bool   isOn()    {       return on;      }

        // PRECONDITION:    must be 'on' for valid
        // action & response
```

```
public bool  valid(int x)
  {    if (!on)    return false;
       if (x < lowBound || x > upBound)
              return false;
       if (x == lowBound || x == upBound)
  {    if (inclusive) count++;
           return inclusive;
       }
       count++;
       return true;
  }


// PRECONDITION: must be 'on' to trigger action
public bool  reset()
  {    if (!on)    return false;
       count = 0;
       return true;
  }


// POSTCONDITION:        state of 'on' inverted
  public bool  toggleOn() { return on = !on; }

// PRECONDITION:  must be 'on' for valid response
  public int getCount()
  {    if (on)    return (int)count;
       return -1;
  }
}
```

The class inRange provides two accessor methods (which would be const in C++): 1) isOn() which is unconditional so that client can determine, without restriction, whether an object is usable or not; 2) get-Count() which is conditional because an inactive object does not support queries and should return an error code. Two mutators are also provided: 1) toggleOn() which is unconditional so that client can invert the active state of an object; 2) reset() which is conditional to support the design decision that an inactive object cannot alter state.

Programming by Contract documents design and client responsibility across method pre and postconditions, and interface, implementation, and class invariants. Only one post condition is evident in the class inRange because only one method changes state. Post conditions describe only an actual or potential state change; they do not describe what the method does: valid(int) may alter count but that increment would not constitute a state change. The simplicity of inRange yields an interface invariant that specifies only expected values for the constructor – all other, relevant details for use are enumerated in the class invariant. Self-documenting code leaves little to be specified for Example B.1.1. Sample class and implementation variants are listed below.

1. **Class invariant for** inRange:

   - encapsulation of range (low <= upper)
   - provisional inclusion of boundary values
   - support for both negative and positive boundary values
   - queries of values in range counted
   - delineate utility relative to state on (or off)
   - describe error response if methods invoked when off

2. **Implementation invariant**:

   - re-ordering of boundary values in constructor
   - logic behind error response

This problem is an abstraction (simplification) of a query or logging process that counts legal inquiries, values that fall within some continuous range (cost, age, quantity, etc.). Our design only tracked queries relative to a specified range and thus supports the **Single Responsibility Principle**: *every object should have a single responsibility that is strictly encapsulated.* The only reason to modify this class would be if the tracking criteria changed. We reuse this class in Section B.3 and examine maintainability in Section B.4.

## B.2 OWNERSHIP: C++ CLASS MEMORY MANAGEMENT

Chapter 2 examined ownership, emphasizing the C++ class designer's responsibility to track heap memory. The design exercise from Chapter 2 was to construct a feeLedger container. Using the classic notion of a ledger, as an archive of financial transactions, the feeLedger type provides functionality to record fees and to identify minimum, maximum, mean, and median values. Since transactions are not removed from a record, deletion of fees is not supported. However, a reset method is provided so that the entire ledger can be erased (comparable to deleting a ledger object but allows the allocated heap memory to be reused). Internal storage is required to determine the median. Minimum and maximum values may be easily updated upon receipt of each incoming value. A running sum is tabulated so that the mean (average) value can quickly calculated.

Expectations for unbounded capacity are common and can be supported by using a vector (or another STL container). However, STL structures often incur a performance cost. Example B.2.1 uses an array to demonstrate direct manipulation of heap memory and the definition of a constructor and destructor. Design rests on encapsulation: the client need not know or care about resources used to hold data; when incoming data would overflow the container, resizing is automatically triggered.

Example B.2.1 highlights internalized memory management responsibilities for C++ classes with suppressed copying. See Section B.3 for support of deep copying in both C++ and C#. Acquisition and release of heap memory is illustrated via the constructor, the destructor and a resize method. A constructor sets the initial capacity of the object, which is set to a default if not specified by the client. C++17 supports in-field initialization of data members. A comparable constructor in legacy C++ code would call reset() to initialize data members. A destructor is needed to deallocate internal heap memory when an object goes out of scope. resize() supports unbounded storage capacity by allocating an array of doubled capacity, copying previously stored values into the new array and deallocating prior storage.

### Example B.2.1 C++ Class with Heap Memory

```
#include <algorithm>

class feeLedger
{    const int  HIGHEST_FEE = 10000;
     unsigned   min = INT_MAX;
     unsigned   max = 0;
     unsigned   sum = 0;
     unsigned   count = 0;
     unsigned   capacity;
     unsigned*  array;
```

```cpp
    // copy suppressed - for full support
    // see Section B.3
    feeLedger(const feeLedger&);
    void operator=(const feeLedger&);

    void resize()
    {    capacity *= 2;
         unsigned* temp = new unsigned[capacity];
         for (int k = 0; k < capacity/2; k++)
              temp[k] = array[k];
         delete[] array;
         array = temp;
    }
public:
    feeLedger(unsigned alloc = 100)
    {    capacity = alloc;
         array = new unsigned[capacity];
    }

    ~feeLedger()             {    delete[] array;    }

    //PRE CONDITION:  x may not be zero
    bool record(unsigned x)
    {    if (x == 0)                 return false;
         if (count == capacity)    resize();
         array[count] = x;
         count++;
         if (x > max)    max = x;
         if (x < min)    min = x;
         sum += x;
         return true;
    }

    void reset()
    {    min = INT_MAX;
         max = sum = count = 0;
    }

    unsigned   getMax()  {    return max;   }

    unsigned   getMin()
    {    if (count == 0)         return 0;
         return min;
    }
```

```
double      getMean()
{    if (count == 0)          return 0;
     return sum/count;
}


unsigned   getMedian()
{    if (count == 0)          return 0;
                              // sort in <algorithm>
     sort(array, array + count);
     return array[count/2];
}
};
```

To reduce unnecessary computation, with the assumption that the median is requested infrequently, stored values are sorted only when the median value is requested. A feeLedger object may be reset, that is, returned (mostly) to its initial state. HIGHEST _ FEE is an arbitrary value, encapsulated as a constant. Should the (language provision of the) maximum integer value be used instead? Should an upper bound on fees be passed to the constructor? Such design decisions should be documented in either the class or implementation invariant.

Programming by Contract details for Example B.2.1 differ from those of Example B.1. Most strikingly, there is little need for pre and postconditions. Why? The capacity of the container is unbounded, with internal resizing. The use of uint in C# and unsigned in C++ documents the exclusion of negative integers (but the client should be aware of coercion, when a negative number is converted to a very large positive). There is no state (such as on/off) that restricts valid use of public functionality. Only one precondition is specified: the non-negative integer passed to record() cannot be zero. If the client does pass a zero, no fee is recorded. Since this singular precondition does not involve object state, the client need not track state and thus, there is no need for post conditions.

1. **Class invariant for** feeLedger:

   - unbounded capacity

   - suppressed copying

   - zero as invalid (non-negative) input fee

   - error response – zero as an error code

2. **Interface invariant**:

- Call by value not supported

- Assignment not supported

3. **Implementation invariant**:

- resizing details – double capacity

- conditional processing of input fees—zero invalid

- logic behind error response

- heap memory is not released in `reset()` for efficiency

This design of `feeLedger` clearly restricts copying while storing incoming values and supporting queries for min, max, mean, and median values. Single Responsibility is upheld as is the principle of **Responsibility Driven Design** (RDD) – *all object responsibilities (functionality) and required data are identified.*

## B.3 COPYING

Chapter 3 examined the complexity of copying objects when encapsulated data members (pointers) address heap memory. Use of the STL obviates the need to track many C++ memory management details. Yet, developers should understand the mechanics of copying for both legacy code and performance customization. Shallow copying, the default in both C# and C++, was sufficient in Example B.1 because no heap memory was allocated inside `inRange`. Copying was suppressed in Example B.2.1, invalidating call by value and assignment for the C++ client. Chapter 3's design exercises serve to contrast C# and C++ approaches to deep copying. The first problem was to redesign `feeLedger` from Chapter 2 to support copying.

**Example B.3.1  C++ Class B.2.2 with Deep Copying**

```
#include <algorithm>

class feeLedger
{    const unsigned  HIGHEST_FEE = 10000;
     unsigned  count = 0;;
     unsigned  capacity;
     unsigned* array = nullptr;
```

```cpp
                                        // deep copying supported
    void copy(const feeLedger& src)
    {   count = src.count;
        capacity = src.capacity;
        array = new unsigned[capacity];
        for (int k = 0; k < capacity; k++)
            array[k] = src.array[k];
    }


    void resize()
    {   capacity *= 2;
        unsigned* temp = new unsigned[capacity];
        for (int k = 0; k < capacity/2; k++)
            temp[k] = array[k];
        delete[] array;
        array = temp;
    }
public:
    feeLedger(unsigned alloc = 100)
    {   capacity = alloc;
        array = new unsigned[capacity];
    }
    ~feeLedger()                    {   delete[] array;  }

    feeLedger(const feeLedger& source)
    {   copy(source);    }

    void operator=(const feeLedger& rhs)
    {   if (this == &rhs)       return;
        delete[] array;
        copy(rhs);
    }


    feeLedger(feeLedger&& source) // move semantics
    {   count = source.count;
        capacity = source.capacity;
        array = source.array;

        source.count = 0;
        source.capacity = 0;
        source.array = 0;
    }
```

```
    void operator=(feeLedger&& rhs)
    {    swap(count, rhs.count); //using std::swap()
         swap(capacity, rhs.capacity);
         swap(array, rhs.array);
    }

    void add(unsigned x)
    {    if (count == capacity)    resize();
         array[count] = x;
         count++;
    }

void clear()              {    count = 0;         }

unsigned   getMin()
{    if (count == 0)          return 0;
     unsigned min = HIGHEST_FEE;
     for (int k = 0; k < count; k++)
         if (array[k] < min)
             min = array[k];
     return min;
}

unsigned   getMax()
{    if (count == 0)          return 0;
     unsigned max = 0;
     for (int k = 0; k < count; k++)
         if (array[k] > max)
             max = array[k];
     return max;
}

float        getMean()
{    if (count == 0)          return 0;
     unsigned sum = 0;
     for (int k = 0; k < count; k++)
         sum += array[k];
     return (float) sum/count;
}
unsigned   getMedian()
{    if (count == 0)           return 0;
                               // sort in <algorithm>
     sort(array, array + count);
     return array[count/2];
}
};
```

Contractual documentation is comparable for the C++ and C# feeLedger designs. The class invariant for both should note capacity, bounded or not, as well as data validity and error responses since these details affect client and class designer. The C++ class invariant records copy semantics when copying is suppressed or the client is aware of internally allocated heap memory. Otherwise, the client may assume that copying proceeds 'normally'. The C# class invariant would record support for shallow and deep copying when so provided.

As a record of design decisions, the implementation invariant should note resizing details, the logic behind defined error responses and data processing. The C++ implementation invariant documents support for move semantics – the client need not know about move semantics since it is a compiler optimization, not a client responsibility. The C# implementation invariant documents internal cloning and type reclamation to buffer the client from type casting responsibilities. Like the Example B.2.1 example, pre and postconditions are not required – there is no state that restricts invocation of public functionality.

1. **Class invariant for** feeLedger:

   - unbounded capacity

   - zero is invalid (non-negative) input fee

   - error response – zero as an error code

2. **Interface invariant**:

   - no extraction of individual items

   - clear() discards all contents

3. **Implementation invariant**:

   - resizing details – double capacity

   - conditional processing of input fees—zero invalid

   - logic behind error response

   - heap memory is not released in clear() for efficiency

   - (C++) move semantics for efficiency

   - (C#) internal casting (retrieval of type from Clone())

The C# design promotes data integrity by providing deep or shallow copies upon request from the client. MemberwiseClone() may yield shallow copying for references; the class designer must use it carefully.

### Example B.3.2  C# Version of Class B.3.1

```
public class feeLedger
{    private const int     highestFee = 10000;
     private uint count;
     private uint capacity;
     private uint[] array;

     private void resize()
     {    capacity *= 2;
          uint[] temp = new uint[capacity];
          for (int k = 0; k < capacity / 2; k++)
               temp[k] = array[k];
          array = temp;
     }

     public feeLedger(uint alloc = 100)
     {    capacity = alloc;
          array = new uint[capacity];
     }
     public feeLedger ShallowCopy()
     {    return (feeLedger) this.MemberwiseClone(); }
     public feeLedger DeepCopy()
     {feeLedger f = (feeLedger)this.MemberwiseClone();
          f.array = new uint[capacity];
          for (int k = 0; k < capacity / 2; k++)
               f.array[k] = array[k];
          return f;
     }

     public void add(uint x)
     {    if (count == capacity) resize();
          array[count] = x;
          count++;
     }

     public void clear()  {      count = 0; }

     public uint getMin()
```

```
{    if (count = = 0) return 0;
     uint min = highestFee;
     for (int k = 0; k < capacity; k++)
          if (array[k] < min) min = array[k];
     return min;
}

public uint getMax()
{    if (count = = 0) return 0;
     uint max = 0;
     for (int k = 0; k < capacity; k++)
          if (array[k] > max) max = array[k];
     return max;
}

public float getMean()
{    if (count == 0) return 0;
     uint sum = 0;
     for (int k = 0; k < capacity; k++)
          sum += array[k];
     return sum / count;
}

public uint getMedian()
{    if (count == 0) return 0;
     Array.Sort(array);
     return array[count / 2];
}
}
```

For Example B.3.2, contractual documentation informs the client of the choice between shallow and deep copying. The implementation invariant must explain the copying technique used since C# supports different approaches. As noted in Chapter 3, when a C# design for deep copying implements the ICloneable interface and overrides Clone(), the client must cast the returned object to the desired type. In contrast, as shown in Example B.3.2, internal casting design frees the client from managing type, that is, reclaiming the appropriate type from a generic reference.

In Example B.3.1, the C++ design promotes efficiency and preserves data integrity by providing deep copying when needed, and allowing the compiler to transfer ownership (via move semantics) when not. In Example B.3.2, the C# client must assess the need for copying, choosing between deep and shallow copying. Both designs adhere to the **Principle of Least Knowledge** (PLK) – also known as the Law of Demeter – *minimize dependencies by limiting knowledge that objects have of each other.* Why? Copying is internalized in each class design: the client need not track implementation details. Since deep copying in C# requires casting, and Example B.3.2 internalizes type reclamation from the generic object, PLK is further supported.

**Example B.3.3  C++ Redesign Using STL Vector**

```cpp
#include <algorithm>
#include <vector>
#include <numeric>
// C++ redesign using vector
class feeLedger
{    const int  HIGHEST_FEE = 10000;
     vector<unsigned>          fees;
public:
     void add(unsigned x) {  fees.push_back(x);      }
     void clear()         {  fees.clear();           }

     unsigned   getMax()
     { if (fees.size() == 0)      return 0;
       return *max_element(fees.begin(),fees.end());
     }

     unsigned   getMin()
     { if (fees.size() == 0)      return 0;
       return *min_element(fees.begin(),fees.end());
     }

     float      getMean()
     { unsigned sum = accumulate(fees.begin(),
                        fees.end(),0);
         return (float) sum/fees.size();
     }

     unsigned   getMedian()
     {    sort(fees.begin(), fees.end());
          return fees[fees.size()/2];
     }
};
```

## B.4 COMPOSITION

Chapter 4 examined composition designs that reflect type dependency and reuse code. Composition yields stability via wrapping, callback via delegates, efficiency via postponed instantiation, and maintainability via Dependency Injection. Internal control stressed variability of ownership, association, cardinality, and lifetime for encapsulated subobjects.

The first exercise was to define class openRange to track integers queried within a specified range, much like inRange from Chapter 1 but with the ability to provide boundary values upon request. Key differences are the: 1) provision of getters to retrieve upper and lower bound values; 2) exclusion of boundary values from the encapsulated range; 3) lack of state control over access (on/off). openRange is a simple wrapper, with complete type dependency on its inRange subobject. The openRange class designer chooses what functionality to echo since composition supports selective accessibility. getCount(), valid(), reset()) are all likely to be inlined so there is no run-time penalty for accessing these wrapped methods. Example B.4.1 demonstrates a C# implementation of wrapper class openRange which operates as a data evaluator, tracking data relative to a specified range and supporting the retrieval of boundary values.

**Example B.4.1  C# Class with Composition for Code Reuse**

```
public class openRange
{   private inRange       reUsed;
    private readonly int lowBound;
    private readonly int upBound;

    // 1) provision of getters
    public int  getLowerBound()    { return lowBound;}
    public int  getUpperBound()    { return upBound;}

    public int  getCount()
    {    return reUsed.getCount(); }

    public bool valid(int x)
    {    return reUsed.valid(x);    }
    public bool reset()  { return reUsed.reset(); }

    public openRange(int   x, int y)
    {    lowBound = x < y? x: y;
```

```
          upBound =   x < y ?  y: x;
          // 2) boundary values always excluded
          // by passing false
          reUsed = new inRange(lowBound,
                   upBound, false);
    }
    // 3) state control of inRange NOT supported
    // => NOT echoed
    //   bool    toggleOn()    and  bool    isOn()
}
```

As defined in Example B.1, inRange operated as a data evaluator, tracking data with respect to a specified range but did not store data or support the retrieval of boundary values. What if the client wishes to retrieve boundary values? Requiring the client to record values passed into a constructor, and retain the connection between a specific object and its instantiation, defeats the abstraction and encapsulation of OOD. Since a persistent object may be used far from its instantiation point, it is difficult for a client to preserve the association between an object and values passed into its constructor. The openRange wrapper thus augments the inRange design, expanding its utility.

1. **Class invariant for** openRange:

   • encapsulation of range

   • boundary values not included in range

   • boundary values revealed upon request

   • queries of values in range counted

2. **Implementation invariant**:

   • Wrap a inRange object to record boundary values

   • on/off state suppressed

   • inRange methods for on/off not echoed

   • record boundary values as passed into constructor

Our wrapper class suppresses the (on/off) state of inRange – a viable design option because the change to on/off is externally triggered. There is no limit to wrapping. openRange could easily be wrapped to add the tracking of the **min**imum, **max**imum, and average (**mean**) incoming values, see Example B.4.2.

**Example B.4.2 C# Class: Two Layers of Wrappers**

```
public class mmmRange
{      private openRange     reUsed;
       private int           min = Int32.MinValue;
       private int           max = Int32.MaxValue;
       private int           sum;

       public mmmRange(int x, int y, bool border = true)
       {    reUsed = new openRange(x, y);              }

       public int getLowerBound()
       { return reUsed.getLowerBound(); }
       public int getUpperBound()
       { return reUsed.getUpperBound(); }

       public int getCount()
       { return reUsed.getCount(); }
       public bool  valid(int x)
       {    if (!reUsed.valid(x))            return false;
            if (x < min)    min = x;
            if (x > max)    max = x;
            sum += x;
            return true;
       }

       public int getMin()          {    return min;   }
       public int getMax()          {    return max;   }
       public double getMean()
       { return sum/reUsed.getCount(); }
}
```

1. **Class invariant for** mmmRange:

   • encapsulation of range

   • boundary values not included in range

   • queries of values in range counted

   • boundary values revealed upon request

2. **Implementation invariant:**

- Wrap a openRange object to track interest in values in range

- Minimum and maximum updated with each check

- openRange methods echoed

Wrapping is a common example of composition and is an effective design technique to reuse, customize, update, or improve code. The intent to hide unstable details minimizes client responsibilities. Double wrapping can do so as well but realize that each additional layering obscures design. What if a need to track median values, along with minimum, maximum, and mean, arises? One could then wrap mmmRange (which wraps openRange which wraps inRange) and add data storage. At some point though, starting over with a clean design becomes more maintainable, if only because software developers must wade through fewer layers of code.

Designs use composition to avoid unnecessary overhead. Temporary objects are often so short-lived that their functionality is not exercised. In which case, it makes sense to postpone instantiation of any internal resources: there is no need to allocate a resource that will not be used. The next design exercise reuses feeLedger in a composition relationship that postpones instantiation. Note that internal construction, whether postponed or not, minimizes the need to define error response(s).

**Example B.4.3  C++ Composition with Postponed Instantiation**

```
class trafficStats
{       unsigned    idLocale;
        unsigned    volumeNS;
        unsigned    volumeEW;
        unsigned    numFines;
            //    suppress copying
        trafficStats(const trafficStats&);
        void operator=(const trafficStats&);
        feeLedger* fines;
    public:
        trafficStats(unsigned id = 10000)
        {    idLocale = id;
             volumeNS = volumeEW = numFines = 0;
             fines = nullptr;
        }
```

```
void count(bool NS)
{       if (NS)      volumeNS++;
        else         volumeEW++;
}

unsigned getNSCarCount()  {  return volumeNS;  }
unsigned getEWCarCount()  {  return volumeEW;  }
unsigned getNumFines()    {  return numFines;  }

void addFine(int ticket)
{       if (!fines)     fines = new feeLedger();
        fines->add(ticket);
        numFines++;
}

void reset()
{       if (fines)      fines->clear();
        volumeNS = volumeEW = 0;
}

float       getAvgFine()
{       if (!fines)     return 0;
        return      fines->getMean();
}

unsigned  getMedianFine()
{       if (!fines)     return 0;
        return      fines->getMedian();
}

~trafficStats() {       if (fines) delete fines;  }
};
```

1. **Class invariant for** trafficStats:

   • track traffic volume and fines associated with intersection

   • copying suppressed

2. **Interface invariant**:

   • Call by value not supported

   • Assignment not supported

3. **Implementation invariant:**

- Encapsulate a `feeLedger` object to record fines

- Return mean and median fine values by echoing `feeLedger` interface

- Postpone internal instantiation of `feeLedger` object

- Reset clears all counts but does not deallocate `feeLedger` object

In the `trafficStats` design, an internally constructed `feeLedger` delegate stores integer values (fines) and retrieves mean and median values. Efficiency was promoted via postponed instantiation because no `feeLedger` delegate was allocated if no fines were recorded. With its simple, limited use of delegates, `trafficStats` did not support replacement. Delegates, and resources in general, may be acquired from the client via Dependency Injection to easily support replacement.

Example B.4.4 illustrates Dependency Injection. The class `cyclic-Data` generates successive values from an encapsulated sequence, advancing, or retreating in the sequence according to state. Constructor Injection sets the primary sequence upon object construction; if the client passes in a null sequence, the object is unusable. Property Injection permits sequence replacement only for `stalled` objects. If the client does not pass a valid sequence upon object instantiation, the object is unusable until a valid sequence is injected. The ability to request a filtered value, skipping specified values, is supported via Method Injection. The sequence containing values to be skipped (filtered) is passed into `getFilteredValue`.

**Example B.4.4  C# Dependency Injection**

```
public class cyclicData
{    private     int[]      seq;
     private     uint       index;
     private     bool       advance = true;
     private     bool       stalled;

     public cyclicData(int[]  sequence)
     {   if (sequence == null)       stalled = true;
         seq = sequence;
     }
```

```
public bool replaceData(int[] jit)
{   if (stalled && jit != null)
    {   seq = jit;
        stalled = false;
        return true;
    }
    return false;
}

public int getNextValue()
{   if (stalled)    return 0;
    uint cycle = seq.Length;
    if (advance)  index = (index + 1) % cycle;
    else          index = (index - 1) % cycle;
    return seq[index];
}

private bool inSkip(int value, int[] skip)
{   for (int k = 0; k < skip.Length; k++)
        if (value == skip[k]) return true;
    return false;
}

public int getFilteredValue(int[] skip)
{   if (skip == null || stalled)    return 0;
    int value = 0;
    bool inSeq = true;
    uint count = 0;
    while (inSeq && count < seq.Length)
    {   value = getNextValue();
        inSeq = inSkip(value, skip);
        count++;
    }
    if (inSeq)              return 0;
    return value;
}

public void setAdvance(bool y) { advance = y; }
}
```

1. **Class invariant for** cyclicData:

   • Encapsulated cyclic sequence must be injected via constructor

   – Null sequence will yield unusable object (stalled)

- Zero is not a legal value in any sequence – used as an error code

- State transition (forward or backward progression) under client control

- No restriction on sequence length

2. **Interface invariant**:

- Sequence of values to skip injected via property

- Error response to null sequence(s) and all values skipped is return of zero

3. **Implementation invariant**:

- `stalled` not reset once initialize in constructor => object unusable until a valid sequence is provided.

Composition designs reuse types internally, with or without the client's knowledge. The wrapper classes in Examples B.4.1 and B.4.2 do not expose the wrapped type to the client. Likewise, Example B.4.3 reused `feeLedger` without the client's knowledge and thus retained complete control over instantiation and state. The final design exercise for Chapter 4 used Dependency Injection, exposing the resource (an integer array of forbidden values) to the client. All these designs followed the **Dependency Inversion Principle** – *Abstractions should not depend on details – details (concrete implementations) should depend on abstractions.* Clients need not track internal resources when hidden, as in the first three designs. Clients may replace resources that conform to a specified interface, as in the fourth design.

## B.5 INHERITANCE

Chapter 5 analyzed inheritance designs and concluded, possibly surprisingly, that code reuse and accessibility are not sufficient justifications for inheritance. Why? Composition also offers code reuse and can control accessibility. Inheritance is easy to define syntactically and thus may be overused. Type extensibility, substitutability, and client use of heterogeneous collections are unassailable reasons to employ inheritance.

The chapter's first design exercise was to define an inheritance hierarchy of generators, where each generator yields individual values from an internal arithmetic sequence: an `arithSeq` object yields the next value from its arithmetic sequence when in 'advance' mode, the previous value when in 'retreat,' and the current value when 'stuck'. The client may request a value as well as alter mode, reset, and query the count of mode changes. Two descendants of `arithSeq` are:

1. `oscillateA` is-a `arithSeq` and so each `oscillateA` object operates like an `arithSeq` object, except that successive values returned from an `oscillateA` object oscillate between negative and positive values.

2. `skipA` is-a `arithSeq` and so each `skipA` object operates like an `arithSeq` object, except that values returned from a `skipA` object reflect the skipping of some number of values – this skip value should be constant but variable from object to object.

Inheritance designs should evaluate parental control and child accessibility. How much should parent classes keep hidden? Do parent classes define their data as: 1) protected; 2) private with no access for descendants; or, 3) private with protected (or public) accessors and mutators as needed? In Example B.5.1, descendant classes do not compute any values: `oscillateA` may change the sign of the value computed by its parent; `skipA` skips some number of values computed by its parent. Consequently, data members for sequence generation may remain private in the parent class. Since the client may alter the state controlling the forward and backward progression of the sequence, descendant classes should be able to see the Boolean controlling advance or retreat. The encapsulated arithmetic sequence varies from object to object so the `arithSeq` constructor should take in $a_1$ and the distance value; descendant constructors should pass these values up to their parent constructor.

### Example B.5.1  C# Inheritance

```
public class arithSeq
{       private     readonly int     a1;
        private     readonly uint    dst;
        private     uint             index;
        protected   uint             maxCount;
        protected   uint             count;
        protected   bool             advance = true;
        protected   bool             stuck;

    public arithSeq(int a, uint d, uint limit = 10000)
        {   a1 = a;
            if (d == 0)     d = 10;
            dst = d;
```

```
            if (limit < 100)            limit = 101;
            maxCount = limit;
    }

    public virtual int getNextValue()
    {   if (!stuck)
        {   if (advance)      index++;
            else              index--;
        }

        count++;
        stuck = (count >= maxCount);
        return a1 + (int)(index*dst);
    }

    public void reset(uint newLimit = 0)
    {   if (newLimit = = 0)  newLimit = 2*maxCount;
        maxCount = newLimit;
        stuck = false;
        index = count = 0;
        advance = true;
    }
    public uint getQueryCount()   { return count; }
    public bool setAdvance(bool y)
    { return advance = y; }
}

public class oscillateA:  arithSeq
{   private     int   prev;

    public oscillateA(int a, uint d,
      uint k = 1000):base(a,d,k)
    { }

    public override int getNextValue()
    {   int value = base.getNextValue();
        if ( ((prev > 0) &&   (value < 0 )) ||
             ((prev < 0) &&   (value > 0 ))   )
            return value;
        return value = -value;
    }
}

public class skipA:  arithSeq
{   private readonly     uint   skip;
```

```
public skipA(int a, uint d,
  uint k = 10000):base(a,d,k)
{    skip = d/2 +2;   }
public override int getNextValue()
{    uint trueCount = count;
     int value = 0;
     for (int k = 0; k < skip;  k++)
          value = base.getNextValue();
     count = ++trueCount;
     return value;
}
}
```

1. **Class invariant for** `arithSeq`:

   - NO inherent restriction on numeric values

   - Values generated from $a_1$ and distance as passed to constructor

     - limit on number of values returned (until `reset`) – default 101

   - Sequence progression (forward or backward) under client control

2. **Interface invariant**:

   - `reset()` restarts sequence generation in forward progression

     - Optionally may pass limit on number of values returned

3. **Implementation invariant**:

   - `stuck` not permanent status => object may be reset and thus indefinitely usable

   - NO `const` or `static` values for sequence

   - NO internal random number generator

   - `getNextValue()` virtual because child classes may alter value returned

   - `count` is protected so that child classes may control variations

   - remaining data members private because child classes

     - do not change implicit arithmetic sequence

     - do not alter state or most bookkeeping detail

`getNextValue()` supports polymorphic behavior. However, since descendant classes do not change the underlying arithmetic sequence or any bookkeeping details, their contractual details are minimal. `skipA` discards some number of values computed by its parent: its implementation invariant must specify whether the skip value is passed to the constructor or is internally generated. A potential inconsistency is that the parent maintains the count of values generated but `skipA` may drive that count up while only yielding one value for the client. To prevent such an error, `skipA` records the parent count and updates its value only upon return.

1. **Class invariant for `skipA`:**

   - see parent `arithSeq` contractual details

   - sequence generation is altered by the intentional discarding of some values

2. **Implementation invariant:**

   - see parent `arithSeq` contractual details

   - number of values to skip is stable, set in the constructor and dependent on `dst`

   - parental control of query count must be circumvented so that discarded values not included

Type extensibility is clearly supported in Example B.5.1. A new child class may be added without affecting existing class or breaking client code. The second design exercise in Chapter 5 was to modify the generator type definition so that, internally, values are filtered. When a generator object yields a value that is in a 'forbidden' set, that value is skipped. Dependency Injection is used to 'inject' the forbidden set into a generator object. Since the key virtual method `getNextValue()` takes no parameters, the forbidden set may not be acquired via Method Injection. Constructor or Property Injection may be used as illustrated below. Again, a problem arises from the parent counting the values generated. `skipA2` may consume multiple values in the quest to find one outside the forbidden set. Again, the child class may record the parent count and update its value only upon return. But there are other difficulties due to hidden data in the parent class. What if the generator becomes stuck? It will return the same value, leading to an infinite loop if that value is in the forbidden set. stuck is private, and there is no getStuck() method. One solution is to call reset() as shown below.

### Example B.5.2  C# Inheritance with Dependency Injection

```
public class skipA2: arithSeq
{     private int[]   discard;

      // Constructor Injection of forbidden set
      public skipA2(int a, uint d,
        uint k = 1000, int[] f = null) :base(a,d,k)
      {    discard = f;      }

      // Property Injection of forbidden set
      public void setDiscard(int[] f) { discard = f; }

      private bool notMember(int value)
      {     for (int k=0; k < discard.Length; k++)
              if (value = = discard[k])   return false;
            return true;
      }

      public override int getNextValue()
      {     uint trueCount = count;
            int value = base.getNextValue();
            bool done = (notMember(value));

            while (!done)
            {    if (stuck)       reset();
                 value = base.getNextValue();
                 done = (notMember(value));
            }
            count = ++trueCount;
            return value;
      }
}
```

Inheritance designs rest on a stable interface, supporting client familiarity with type. The designs illustrated here adhere to the **Liskov Substitutability Principle (LSP)** – *a child object can stand in (substitute) for a parent object*. Thus, the client has access to a variety of subtypes, and their variant behavior as supported by virtual methods. Key benefits of good inheritance design are the use of heterogeneous collections, stability of interface, and type extensibility.

# Appendix C: Comparative Design Exercises

**Design Choices and OOD Principles for Chapters:**

| | |
|---|---|
| 6) Is-a vs Has-a | *Composite Principle* and *Open Closed Principle* |
| 7) Design Longevity | *Don't Repeat Yourself* |
| 8) Operator overloading | *Program to Interface Not Implementation* |

## C.1 COMPOSITION VERSUS INHERITANCE

Chapter 6 contrasted code reuse via composition versus inheritance. The first exercise starts with an existing type definition for which there is little difference between code reuse via composition or inheritance. Subsequent designs illustrate significant impact of relationship choice on the consistency, efficiency, and flexibility of code reuse.

Class design was highlighted in Chapters 1–3. Recall the feeLedger class from Example B.2.1 that stored data and tracked minimum, maximum, median, and mean values. Consider defining a new type that stores only filtered data values. Should feeLedger be reused via composition or inheritance? That is, should the filtering process be added in a wrapper or a child class? The two approaches are compared below.

A composition design wraps feeLedger and adds a filter to store only values that fall within a specified range. Reuse of both the feeLedger class and the inRange class (from Example B.1.1) reduces development overhead: an incoming value is tested to see if it is a valid fee (within range); otherwise, it is not pass onto the encapsulated feeLedger object. Example C.1.1 shows a composition design for filterLedger.

Example C.1.2 shows an inheritance design. Since the inRange delegate is optional, it is allocated on the heap in setFilter(). Although the client is unaware of implementation details, the client must request a filter and pass in appropriate bounds.

If filtering is not activated, there is no need to construct a delegate: instantiation should be postponed to avoid unwarranted overhead. Thus, existence checks must be included in all methods that may use the inRange delegate. setFilter() is <u>not</u> a true example of Property (Setter) injection because the client does not inject an inRange object. inRange reuse is internalized and hidden from the client. inRange could be replaced by outRange without any client knowledge. Given the dependency of filterLedger on feeLedger, a feeLedger subobject is instantiated upon construction of a filterLedger object.

### Example C.1.1  C# Reuse via Composition

```
// filterLedger wraps feeLedger;
// conditionally, uses inRange
public class filterLedger
// zero initialized
{    private     inRange     tracker;
     private     feeLedger   book;

     public filterLedger(unsigned alloc = 100)
     {    book = new feeLedger(alloc);   }

     public void  setFilter(int x, int y)
     {    tracker = new inRange(x,y);                 }
     public void add(uint x)
     {    if (tracker != null &&
               !tracker.valid((int) x)) return;
          book.add(x);
     }

     public void     clear()     {  book.clear();  }
     public uint     getMin()
     {  return book.getMin();  }
     public uint     getMax()
     {  return book.getMax();  }
```

```
public double    getMean()
{   return book.getMean();}
}
```

For both composition and inheritance designs, the class invariant should describe the option of filtering fees. The implementation invariant must explain the choice of an inRange object to *internally* check the validity of a fee – the client need not know about the inRange delegate so this detail is not published in the class or interface invariants. The interface invariant reminds the client that reset() restores the feeLedger object to its initial state, discarding all stored values.

### Example C.1.2  C# Reuse via Inheritance

```
// child class filterInheritLedger
// conditionally, uses inRange
public class filterInheritLedger: feeLedger
// zero initialized
{    private    inRange    tracker;
     public filterInheritLedger(uint alloc = 100)
           :base(alloc) {}

     public void  setFilter(int x, int y)
     {    tracker = new inRange(x,y);                   }

     public void add(uint x)
     {    if (tracker != null &&
              !tracker.valid((int) x)) return;
          base.add(x);
     }

     public void clear()
     {    tracker = null;
          base.clear();
     }
}
```

Examples C.1.1 and C.1.2 illustrate that reuse via inheritance versus composition yields little difference. There is no need to postpone instantiation, vary cardinality, or replace the feeLedger component, so composition is not strongly warranted. None of the `feeLedger` functionality (get-Min(), getMax(), getMean()) is dynamically bound so inheritance is also not strongly warranted. Comparable C++ implementations of both designs would yield little difference, other than the definition of copy semantics for the optional `inRange` delegate which would be held on the heap.

The next exercise was to design two classes, `factor` and `twoFactor`. The `factor` class provides functionality to check divisibility of an input value by one encapsulated *factor* and maintain a count of successful queries. The `twoFactor` class provides functionality to check divisibility of an input value by *two factors* and maintain a count of successful queries. The type definition of `twoFactor` should reuse `factor`. Should reuse be achieved through inheritance or composition?

### Example C.1.3  Reuse via Inheritance

```
/* factor:
      encapsulates one non-zero value, j
      divide(z)determines if z is
        evenly divisible by j
      counts the number values received
        that are evenly divisible by j
twoFactor:
      supports all the public capabilities
        of a factor object
      tracks two non-zero values, j and k
      divide(z) determines if z is evenly
        divisible by j and k
      e.g., if j=2 and k=5, then 30 is evenly
        divisible by j and k but 21 is not */

public class factor
{       private int  j;
        private uint count;

        public factor(int j)
        {       if (j == 0)  j = 2;
                this.j = j;
        }

        public bool divide(int  z)
        {       if (z % j == 0)     return false;
```

```
            count++;
            return true;
    }

    public uint Count    {   get;   }
    public void reset() {   count = 0;          }
}

public class twoFactorInherit: factor
{       private int  k;

    public twoFactorInherit
            (int j, int k): base(j)
    {       if (k == 0)   k = 1;
            this.k = k;
    }

    public bool divide(int   z)
    {       if (z % k == 0)         return false;
            return base.divide(z);
    }

}
```

Example C.1.3 shows reuse via inheritance but the parent component tracks divisibility for only one factor, j, so twoFactorInherit uses an additional int to track the divisibility of k. The child class requires its own count because the parent queries do not accurately reflect divisibility by <u>both</u> j and k. Example C.1.4 revises twoFactorInherit. Using both composition and inheritance, twoFactorMixed replaces the int data member with a factor data member, expanding reuse of the factor class. The code is cleaner than that of Example C.1.3, though it is a mixed design. Why use inheritance at all? twoFactorComposed reuses the factor class using only composition: one factor for j, a second factor for k.

**Example C.1.4 Reuse via Composition**

```
public class twoFactorMixed: factor
{       private factor             k;

    public twoFactorMixed
            (int j, int k): base(j)
    {       this.k = new factor(k);      }
```

```
public bool divide(int  z)
{       if ( base.divide(z) && k.divide(z))
        {       count++;
                return true;
        }
        return false;
}

public uint Count    { get; }

public void reset()
{       count = 0;
        base.reset();
        k.reset();
}
}

public class twoFactorComposed
{       private factor      j;
        private factor      k;
        private uint        count;

        public twoFactorComposed(int j, int k)
        {       this.j = new factor(j);
                this.k = new factor(k);
        }

        public bool divide(int  z)
        {       if ( j.divide(z) && k.divide(z))
                {       count++;
                        return true;
                }
                return false;
        }

        public uint Count    { get; }

        public void reset()
        {       count = 0;
                j.reset();
                k.reset();
        }
}
```

What if one wants to track queries that are divisible by three factors? Would it be reasonable to reuse either class in Example C.1.4 and add another factor data member? Incremental expansion of a type definition is not maintainable. If cardinality seems likely to expand, then use an expandable container. An array of factors generalizes design, streamlining support for multiple factors. The client can specify the required factors via the constructor. Example C.1.5 illustrates this design.

**Example C.1.5 Streamlined Reuse via Composition**

```
public class manyFactorC
{       private readonly factor[]   array;
        private uint                numFactors;
        private uint                count;

        public manyFactorC(int[] divisors)
        {       numFactors = (uint) divisors.length;
                array = new factor[numFactors];
                for (int k = 0; k < numFactors; k++)
                    array[k] = new factor(divisors[k]);
        }

        public bool divideByAll(int   z)
        {       bool divisible = true;
                uint index = 0;
                while (divisible && index <
                numFactors)
                {       divisible = array[index].
                        divide(z);
                        index++;
                }
                if (divisible) count++;
                return divisible;
        }

        public uint Count   { get; }

        public void reset()
        {       count = 0;
                for (int k = 0; k < numFactors; k++)
                        array[k].reset();
        }
}
```

The third exercise in Chapter 6 is to design an amorph inheritance hierarchy to model variant behavior. All amorph objects move across a grid (x, y). Initially, all objects start with the same brightness and size. Each descendant varies a different, inherited feature. dimmer varies its brightness but only if it has moved; swatch changes its color but only if it has not moved more than some number of times; nimble can change its size but only relative to the number of times it has moved. We present two solutions, both define a stable interface and use polymorphism (and thus, inheritance).

**Example C.1.6 Inheritance with Polymorphic change()**

```
public enum Color { blue, green, yellow, red,
                    magenta, purple }

public class amorph
{       private     uint    x;
        private     uint    y;
        protected   float   bright = 10;
        protected   Color   c = Color.blue;
        protected   uint    size = 10;
        protected   uint    numMoves;

        public amorph(uint x, uint y, Color c)
        {       this.x = x;   this.y = y;
                this.c = c;    }

        public void move(uint x, uint y)
        {       numMoves++;
                this.x = x;   this.y = y;
        }

        public virtual bool change(int v)
        // default behavior is NOP
        {       return false; }
}

public class dimmer: amorph
{       public dimmer(uint x, uint y, Color c):
            base(x,y,c)  {}
```

```
        public override bool change(int update)
        {       if (numMoves == 0)        return false;
                bright += update % 10;
                return true;
        }
}

public class swatch: amorph
{       private        uint   limitMoves;

        public swatch(uint x, uint y, Color c):
          base(x,y,c)
        {       limitMoves = x*y;       }

        public override bool change(int update)
        {       if (numMoves >= limitMoves)
                  return false;
                c = (Color) update;
                return true;
        }
}

public class nimble: amorph
{       public nimble(uint x, uint y,   Color c):
          base(x,y,c) {}

        public override bool change(int update)
        {   if (numMoves < update)    return false;
            size += (uint)(numMoves + update) % size;
            return true;
        }
}
```

Example C.1.6 uses a straight-forward inheritance design with a polymorphic method change() that serves to 'catch all' variant behavior. change() ensures that the interface of amorph is stable. Example C.1.7 sketches a different approach using the Template Method. The primary method change() is statically bound and not overridden by descendant classes. Instead, each descendant defines the protected method modify() that is abstract in the base class. change() implements common pre-processing and then triggers the variant behavior by invoking the dynamically bound modify(). The Template Method inverts control: the parent calls the child.

### Example C.1.7 Template Method for Stable Interface

```
public class amorph
{        private      uint   x;
         private      uint   y;
         protected    float  bright = 10;
         protected    Color  c = Color.blue;
         protected    uint   size = 10;
         protected    uint   numMoves;

         protected virtual bool modify(int v)
         // default behavior is NOP
         {      return false; }

         public amorph(uint x, uint y, Color c)
         {      this.x = x;   this.y = y;
                this.c = c;     }

         public void move(uint x, uint y)
         {      numMoves++;   this.x = x;
                this.y = y;     }

         public bool change(int update)
         {      // common pre-process
                return modify(update);
         }
}

public class dimmer: amorph
{        public dimmer(uint x, uint y, Color c):
           base(x,y,c){}

         protected override bool modify(int update)
         {      if (numMoves == 0)   return false;
                bright += update % 10;
                return true;
         }
}
```

```
public class swatch: amorph
{       private       uint    limitMoves;

        public swatch(uint x, uint y, Color c):
          base(x,y,c)
        {       limitMoves = x*y;       }

        protected override bool modify(int update)
        {       if (numMoves >= limitMoves)
                  return false;
                c = (Color) update;
                return true;
        }
}

public class nimble: amorph
{       public nimble(uint x, uint y, Color c):
          base(x,y,c) {}

        protected override bool modify(int update)
        {       if (numMoves < update)
                  return false;
                size += (uint)(numMoves + update)
                                          % size;
                return true;
        }
}
```

The final exercise in Chapter 6 directed the reuse of the amorph inheritance hierarchy to define shapeShifter: a type that can be switch from one amorph subtype to another. This exercise prepares the reader for the disassembler example of Chapter 7 – where the client (the engineer writing disassembler code) changes subtypes with each failed guess. The type shapeShifter likewise supports subtype change but internalizes changing subtypes via a polymorphic delegate. Example C.1.8 displays this composition design – variant behavior rests on the delegate amorph inheritance hierarchy.

### Example C.1.8 Replaceable Polymorphic Delegate for Varying Behavior

```
public class shapeShifter
{       private amorph       myDelegate;
```

```
public shapeShifter(uint x, uint y, Color c)
{      myDelegate = new nimble(x, y, c); }

public bool move(uint x, uint y)
{      if (myDelegate == null) return false;
       myDelegate.move(x,y);
       return true;
}

public bool change(int update)
{      if (myDelegate == null) return false;
       return myDelegate.change(update);
}

public bool morph(amorph newSkin)
{      if (newSkin == null) return false;
       myDelegate = newSkin;
       return true;
}
}
```

The sample exercises in this first section of Appendix C contrast the use of composition to inheritance. The first two exercises, Examples C.1.1–C.1.5, display the advantages of composition: postponed instantiation, dependency injection, transfer of ownership, replacement, and selective suppression and/ or echoing of interface. These designs sustain internal control of delegates, adhering to the **Composite Principle**: *prefer composition over inheritance*.

The third exercise, Examples C.1.6–C.1.7, displayed the primary motive to employ inheritance: polymorphic behavior as triggered through a stable, parent interface. The last exercise, Example C.1.8 illustrated the use of composition alongside inheritance via polymorphic delegates. All three examples show the benefits of substitutability, support for heterogeneous collections, and maintainability due to type extensibility, demonstrate the relevance of the **Open Closed Principle (OCP)**: *a class should be open for extension but closed to modification*.

These examples illustrate that different designs are valid and reasonable and that different priorities encourage the choice of one over the other. Code is reused either way. Accessibility constraints may be circumvented so protected data members and functionality need not dictate design choices. Ultimately, the class designer must determine the desired effects of design choices as well as the intended use and longevity.

## C.2  DESIGN LONGEVITY

Chapter 7 exercises focused on design longevity. The first exercise was to construct a hierarchy of *maps*, where each *map* object encapsulates a 2D map, restricting data values, and supporting the following functionality:

1. populate an encapsulated 2D array of integers, with special border values

2. conditionally overwrite the value in row x, column y

3. 'freeze' a map location; frozen locations may not be overwritten

4. for a specified row, return the minimum or maximum value to client

5. for a specified column, return the minimum or maximum value to client

An abstract class requires inheritance to complete a type definition. A base class establishes a common interface to be supported by all descendants, and may or may not initialize protected data used by descendants. Virtual methods signify what public behavior is expected to be redefined by descendants. The abstract base class in Example C.2.1 relies on descendant classes to determine data validity and thus does not overwrite C#'s zero initialization. Each map subtype determines its own restrictions on data values. Descendant classes are:

1. *modMap* is-a *map* that encapsulates a 2-digit 'mod' value m, used to ensure that all values in the map are evenly divisible by m. *modMap* objects will not freeze any values in a row or column whose indices are evenly divisible by m.

2. *uniqueMap* is-a *map* that holds no duplicates and will not freeze values.

Examples C.2.1 and C.2.2 illustrate: 1) dynamic binding reliant on the base class interface; 2) internal dynamic binding within the Template Method pattern; 3) exported (client) responsibility for type specialization; 4) support of client type extraction via type identification in the base class. The first two options rest on a stable public interface and are maintainable designs as long as the base class interface is sufficient. The last two options arise from the extension of an inherited public interface. When public methods are either missing from the base class interface or are not accessible via a base class handle, the client must perform type checking 'manually'. The given sample code suggests that contractual documentation notify clients that invalid row or column indices will cause thrown exceptions.

The Template Method pattern rests on a base class public non-virtual (template) method that provides variant behavior through internal calls to protected methods. The template method determines the basic steps for fulfilling the required functionality. Tagged as a (protected) virtual method, each step may be abstract or defined with default behavior in the base class. Descendant classes may customize the implementation of each step. Example C.2.2 defines the non-virtual method populate() which calls protected, virtual method isValid(). When a client calls populate(), through a base or derived handle, the implicit parameter, the this pointer, yields type information (map, modMap, or unique-Map). The internal call to protected virtual method isValid() yields the variant descendant behavior.

This design, like the previous one, rests on a stable interface. Descendants must work within the confines of the base class template method. For example, isValid() provides variant filtering but is only indirectly accessible through populate(). Descendant classes may alter the behavior of the protected steps but cannot change the order of their execution. The implementation invariant should specify the initial or core functionality provided by the base class and any protected data needed by descendants. Contractual details should include constraints or state changes associated with protected virtual methods. The Template Method design pattern is said to portray the Hollywood principle – don't call us, we'll call you.

### Example C.2.1 Descendants Customize Data Validity

```
public abstract class map
{       protected      int[,]          array;
        protected      bool[,]         frozen;

        private static Random rand = new Random();
        public map(uint x = 100, uint y = 100)
        {       if (x < 10)   x = 100;
                if (y < 10)   y = 100;
                array = new int[x,y];
                frozen = new bool[x,y];
        }
```

```
protected abstract bool isValid(int v);
protected abstract void setBorders();

protected void setBorders(int value)
{       int row = array.GetLength(0);
        int column = array.GetLength(1);
        for (int r = 0; r < row; r++)
        {       array[r,0] = value;
                array[r,column-1] = value;
        }
        for (int c = 1; c < column-1; c++)
        {       array[0,c] = value;
                array[row-1,c] = value;
        }
}


public virtual bool freeze(uint r, uint c)
{       return frozen[r,c] = true;  }

public void populate() // Template Method
{       int row = array.GetLength(0);
        int column = array.GetLength(1);
        for (int r = 1; r < row-1; r++)
                for (int c = 1;
                c < column-1; c++)
                {       int value = 0;
                        bool unverified = true;
                        while (unverified)
                        {       rand.Next(value);
                                unverified =
                                !isValid(value);
                        }
                        array[r, c] = value;
                }
        setBorders();
}

public bool replace(uint row,
uint col, int value)
{   if (!isValid(value) || frozen[row,col])
                return false;
        array[row, col] = value;
        return true;
```

```csharp
        }

        public int getMinCol(uint column)
        {       int min = array[0,column];
                int row = array.GetLength(0);
                for (int r = 1; r < row; r++)
                        if (min > array[r,column])
                                min = array[r,column];
                return min;
        }

        public int getMaxRow(uint row)
        {       int max = array[row,0];
                int column = array.GetLength(1);
                for (int c = 1; c < column; c++)
                        if (max < array[row,c])
                                max = array[row,c];
                return max;
        }
        //  getMaxCol and getMinRow similar
        //  to above methods
}

public class modMap: map
{       protected uint        m;

        public modMap(uint x, uint y): base(x,y)
        {       m = x*y;
                populate();
        }

        protected override bool isValid(int value)
        {       return value % m == 0;        }

        protected override void setBorders()
        {       setBorders((int) m);          }

        public override bool freeze(uint r, uint c)
        {       if ( r % m == 0 ||
                        r > array.GetLength(0))
                return false;
                if ( c % m == 0 ||
                        c > array.GetLength(1))
```

```
            return false;
            return frozen[r,c] = true;
        }
}


public class uniqueMap: map
{       protected override bool isValid(int v)
        {       bool duplicate = false;
                int row = array.GetLength(0);
                int column = array.GetLength(1);
                for (int j = 0; j < row; j++)
                    for (int k = 0; k < column; k++)
                        if (v == array[j,k])
                            return false;
                return true;
        }

        public uniqueMap(uint x, uint y): base(x,y)
        {       populate();    }

        public override bool freeze(uint r, uint c)
        {       return false;      }

        protected override void setBorders()
        {       setBorders(-1);    }
}
```

Example C.2.1 illustrates an effective inheritance design: the base class defines public virtual methods for core functionality with the expectation that descendant classes will augment or replace inherited functionality. Dynamic binding ensures that clients may access variant behavior of subtypes via base class handles. The implementation invariant should specify the initial or core functionality provided by the base class method and any protected data or functionality needed by descendants.

The map hierarchy design appears stable. Type extensions though are uncontrolled and are not guaranteed to be fully utilized through the base class interface. Consider the addition of subtype primeCutMap that stores only non-primes numbers, bounds the number of possible replacements, and freezes no more than half the values in the array. Two of these three design variants are accommodated by the ability to override isValid()and freeze(). However, replace() is statically bound.

Consider also the addition of subtype thawMap that has no restrictions on the data values stored and supports the unfreezing of frozen values. The definition and use of these new subtypes are sketched in Example C.2.2. Recall that the compiler uses the handle type to verify that an invoked method is defined publicly in that class. The vtab used for resolving a dynamic call is identified through the this pointer at run-time. Hence, if a heterogeneous collection is typed to the map class, the compiler looks at the map interface for thaw() and generates a compile-time error upon not finding it. Neither the redefined replace() nor the extended method thaw() are accessible via heterogeneous collections, requiring the client to extract type.

**Example C.2.2  Descendant Types Not Fully Supported by Base Interface**

```
public class primeCutMap: map
{       protected    uint    numReplace;
        protected    uint    limitReplace;

        public primeCutMap(uint x, uint y): base(x,y)
        {       populate();
                limitReplace = (uint) array[0,0];
        }

        protected override void setBorders()
        {       setBorders(1);          }

        protected override bool isValid(int value)
        {       return value.IsPrime();         }

        public override bool freeze(uint r, uint c)
        {       uint count = 0;
                int row = array.GetLength(0);
                int column = array.GetLength(1);
                for (int j = 0; j < row; j++)
                        for (int k = 0; k < column; k++)
                                if (frozen[j,k]) count++;
                if (count > row*column/2)
                        return false;
                return frozen[r,c] = true;
        }
```

```
      public new bool replace(uint row, uint col,
      int value)
      {       numReplace++;
              if (numReplace > limitReplace)
                  return false;
              return base.replace(row, col, value);
      }
}

public class thawMap: map
{       public thawMap(uint x, uint y): base(x,y)
        {       populate();   }

        protected override void setBorders()
        {       return;         }

        protected override bool isValid(int v)
        {       return true; }

        public void thaw(uint r, uint c)
        {       frozen[r,c] = false;            }
}
```

Example C.2.2 displays the effects of an inadequate interface: the client must check type. Extension of a parent class may be considered 'pure inheritance' since the parent class type has not been undermined. A client may use any of the types directly, triggering behavior as associated with each subtype definition. However, consistent usage via a heterogeneous collection is undermined. No client can access the extended interface of a derived class directly through a base class handle. The client must extract type from base class handle at run-time and then invoke the desired (extended) functionality. This design is NOT maintainable or extensible since the addition of any new type to the class hierarchy requires the client to modify code that uses type extraction.

Design inadequacies may arise from insufficient consistency, specifically binding, and complementary functionality. Since the abstract map class expected descendant classes to customize the extent of data alteration (freeze()) then the associated action of data alteration (replace()) should have also been virtual. Similarly, if dynamic change in a particular direction or form is supported, then the class designer should consider complementary (or rollback) functionality. In Example C.2.1, given the ability to freeze data, at arbitrary times, provision of the complementary ability to unfreeze would have yielded a more complete design.

The last design exercise was to reuse two defined types of markers, each of which moves across a two-dimensional grid. An inchworm crawls along a vertical or horizontal line, marking all cells in its path. A leap-Frog jumps from one cell to another along a diagonal, marking only the destination cell. To define a leapWorm class that crawls along a diagonal, reusing these existing types, viable design options are: 1) multiple inheritance (C++); 2) single inheritance from one parent alongside composition (the second parent is subordinated to a data member); and 3) double composition with both parents subordinated to data members.

As noted in Chapter 7, redundancy occurs when reused types have overlapping data, a difficulty not easily resolved by the compiler. Both inchworm and leapFrog move across a grid, marking cells. Each type uses a grid and supports movement. Overlapping data and functionality is clear. Our first design defines both inchworm and leapFrog as descendants of a gridMarker class, with leapWorm then inheriting directly from both inchWorm and leapFrog, reflecting diamond inheritance. Wasted space and initialization overhead would be ill effects because child classes use only the leapFrog grid. Virtual inheritance instructs the compiler to suppress redundant base class components. Data inconsistency does not arise because only the leap-Frog maze in used for movement.

A client may invoke any public inchWorm or leapFrog functionality through a leapWorm object in Example C.2.3. To resolve ambiguity due to inheritance of the same method from both parents, leapWorm must override move(k).

### Example C.2.3  Multiple Inheritance => resolve ambiguity

```
class gridMarker
{    protected:
          Maze           m;
          unsigned       x;
          unsigned       y;
          unsigned       numMoves;
    public:
          gridMarker(unsigned a = 100, unsigned
            b = 100): m(a,b)
          {        x = y = numMoves = 0;        }
```

```
        virtual bool move(unsigned k) = 0;
};

class inchworm: public virtual gridMarker
{   protected:
        unsigned      direction;        // N, E, S, W
        bool          awake;

        bool updateGrid(unsigned k)
        {      bool inMaze = true;
               switch (direction = = 1)
               { case 1:  inMaze = m.markCol(x, x-k);
                          if (inMaze) x-=k;
                          break;
                     case 2: inMaze = m.markRow(y, y-k);
                             if (inMaze) y-=k;
                             break;
                     case 3: inMaze = m.markCol(x, x+k);
                             if (inMaze) x+=k;
                             break;
                  default: inMaze = m.markRow(y, y+k);
                           if (inMaze) y+=k;              }
               }
               return awake = inMaze;
        }
  public:
        inchWorm(int a = 100, int b = 100):
          gridMarker(a,b)
        {      awake = true;
               direction = 1;
        }

        virtual bool move(unsigned k)
        {      if (!awake)           return false;
               numMoves++;
               updateGrid(k);
        }

        virtual void changeDirection(unsigned t)
        {      direction = (t % 4) + 1;    }

        virtual void reset()
        {      m.clear();
```

```
                awake = true;
                direction = 1;
                x = y = numMoves = 0;
        }
};

class leapFrog: public virtual gridMarker
{   protected:
        unsigned      diagonal;      // SW, NE, NW, SE
        bool          alive;

        bool updateGrid(unsigned k)
        {   bool inMaze = true;
            switch (diagonal = = 1)
            {   case 1:   inMaze = m.markCell(x-k,y-k);
                          if (inMaze) { x-=k; y-=k; }
                          break;
                case 2:   m.markCell(x+k,y+k);
                          if (inMaze) { x+=k; y+=k; }
                          break;
                case 3:   m.markCell(x-k,y+k);
                          if (inMaze) { x-=k; y+=k; }
                          break;
                default:  m.markCell(x+k,y-k);
                          if (inMaze) { x+=k; y-=k; }
            }
            return alive = inMaze;
        }
public:
        leapFrog (int a = 0, int b = 0):
          gridMarker(a, b)
        {       alive = true;
                diagonal = 2;
        }

        virtual bool move(unsigned k)
        {       if (!alive)              return false;
                numMoves++;
                updateGrid(k);
        }

        virtual void reset()
```

```
                 { // NOP: frog leaping out of bounds
                   // cannot be reset       }


           virtual void changeDirection(unsigned t)
           {        diagonal = (t % 4) + 1;        }
};

// child class must override overlapping
// move(k) method
class leapWorm: public inchWorm, public leapFrog
{    ...
     public:
           leapWorm(int a, int b):
           inchWorm(a, b), leapFrog(a, b)
           {        ...        }
           bool move(unsigned k)
           {        if (!leapFrog::alive) return false;
                    leapfrog::numMoves++;
                    bool inMaze = true;
                    unsigned numCells = 1;
                    while (inMaze && numCells < k)
                      inMaze = leapfrog::move(numCells++);
                    return alive = inMaze;
           }

           virtual void reset()
           {        m.clear();
                    alive = true;
                    diagonal = 1;
                    x = y = numMoves = 0;
           }

           virtual void changeDirection(unsigned t)
           {        diagonal = (t % 4) + 1;        }
};
```

The sample exercises in this section of Appendix C highlight impediments to and support for sustainable design. The first exercise defines an inheritance hierarchy with and interface presumed to be stable. Subsequent exercises illustrate that this interface is insufficient, forcing client type checking. Type extraction (for dynamic binding and interface extension) yields unmaintainable code. Clients and/or class designers must add compensatory code, replicating type checking that the compiler should provide.

The last exercise motivates a multiple inheritance design and a feasible C++ solution is provided. A comparable C# solution must use composition to mimic at least one parent. With or without composition, the class designer must resolve overlapping interfaces.

**DRY (Don't Repeat Yourself)** is the OOD principle associated with Chapter 8. Through either inheritance or composition, type (class) definitions may be reused, effectively reducing development time and effort. Design longevity is most easily supported through stable interfaces. The use of abstract classes and interfaces support the design of sustainable code.

## C.3 OPERATOR OVERLOADING

The design exercise from Chapter 8 was to transform the C# amplify class, defined in Example 8.7, to a C++ implementation. Since C++ more extensively supports operator overloading, this redesign is not trivial. Potentially, many more operators may be overloaded. Additionally, the manner of overloading varies in C++. In particular, all forms of assignment must be explicitly overloaded. Implementation of pre and post increment (decrement) also differs.

### Example C.3.1 Operators for C++ Class

```
class amplify
{     unsigned    scale;
      bool        on = true;
   public:
      amplify(unsigned amp = 1)
      { scale = amp % 100;   }

      bool isOn() const      { return on;         }
      bool toggleOn() const { return  on = !on; }

      double increase(double x)
      {     if (!on)    return 0.0;
            return      x + (x*scale/100);
      }

      amplify operator+(amplify b)
```

```
{    amplify    local(this->scale + b.scale);
     return local;
}

amplify operator+(unsigned b)
{    amplify    local(this->scale + b);
     return local;
}

amplify& operator+=(amplify b)
{    scale += b.scale;
     return *this;
}

amplify& operator+=(unsigned b)
{    scale += b;
     return *this;
}

// pre ++
amplify operator++()
{    scale = (scale + 1) % 100;
     return *this;
}

// post ++
amplify operator++(int throwAway)
{    amplify local(this->scale);
     scale = (scale + 1) % 100;
     return local;
}

bool operator==(amplify b)
{    return this->scale == b.scale; }

bool operator!=(amplify b)
{    return this->scale != b.scale; }

bool operator<(amplify b)
{    return this->scale < b.scale; }

bool operator>( amplify b)
{    return this->scale > b.scale; }
```

```
       bool operator<=(amplify b)
       {    return this->scale <= b.scale; }

       bool operator>=( amplify b)
       {    return this->scale >= b.scale; }
};
// GLOBAL function in. cpp file
amplify operator+(unsigned a, amplify b)
{      return b + a;   }
```

Operator overloading elevates abstraction and improves readability. Pragmatically, operator overloading supports instantiation of user-defined types in generic algorithms (and containers). Since clients expect consistency, class designers should overload all operators associated with a particular action. In either C# or C++, overloading addition may be a challenging design endeavor. C# enforces paired overloading of comparison operators.

Incorporating operator overloading in a class design stresses the importance of a type interface and hence the associated principle **Program to Interface Not Implementation** (PINI). Adherence to this principle promotes low coupling because the client is not tied to implementation details. Implementation may then change without impacting the client. PINI is particularly relevant to operator overloading because the defined type (class) may be treated like a primitive.

# Glossary

**Abstract class:** is a class definition that is not fully implemented; one or more class methods are undefined with the result that no objects can be instantiated from the class.

**Abstract Data Type (ADT):** is a type definition separated into an interface and an implementation. For example, a stack is an ADT that provides a LIFO (last-in first-out) ordering of data; implementation details of the stack container are not relevant to its use.

**Abstraction:** is the separation of conceptual information from implementation details. For example, a variable name is an abstraction of a memory location, a class interface is an abstraction of its functionality, and a flowchart is an abstraction of control flow.

**Accessor:** is a class method that accesses encapsulated (hidden) data internal to the class. Such methods typically return data by value.

**Ad hoc polymorphism:** refers to function overloading: two or more functions use the same name but can be distinguished by their function signatures (number, order and type of passed parameters).

**Aggregation:** is a form of object composition where the composing object usually contains multiple subobjects but may not necessarily own these subobjects (which do not typically provide functionality to the composing object).

**Aliasing:** occurs when two or more handles (variables) reference the same memory location. Call by reference, for example, sets up an alias between formal and the actual parameters. Aliasing may be used for efficiency since it allows data to be shared, and thus avoids copying. However, aliases must be tracked carefully for data integrity.

**Ambiguity:** denotes a lack of precision that confounds analysis. Compilers cannot handle ambiguity. For example, in a multiple inheritance relationship, when two parent classes define the same named function, it is unclear which method is invoked through a child class object. The compiler cannot resolve such an ambiguous call, so the class designer must resolve the ambiguity by redefining the method (which can simply redirect the call).

**Assembly language:** is a computer language tied to the processor on which it runs and is one step up from machine level code. An assembler translates assembly language code into an executable form (machine code).

**Association:** is the manner in which two or more variables are related, which may be flexible or fixed. For example, a derived class object has a permanent, fixed association with its parent component.

**Base class:** is the topmost (or original ancestor) in a class hierarchy. In a single inheritance relationship, the parent class could also be referred to as the base class. This older term is more often associated with C++ than C# or Java.

**Caching:** is the storage of frequently accessed data so that it can be retrieved quickly. Modern processors have on-chip caches. Programmers may design their own caches, in order to avoid the overhead of memory access, but should do so with care due to the difficulty of ensuring data integrity with two or more copies of the same piece of data.

**Call by Value:** is a parameter passing mode that is considered secure but inefficient. Local memory is allocated and initialized for values passed in and/or out. Thus, local modification of data should not affect external data values. References and pointers undermine the security of call by value.

**Call by Reference:** is a parameter passing mode that is considered efficient but insecure. No local memory is allocated for values passed in and/or out. Instead, formal parameters are aliased with actual argument, thereby avoiding the overhead of data allocation and initialization. Local modification of data does affect external values since memory is shared.

**Cardinality:** is a measure of the number of items in a set, and, in OOD, reflects the number of subobjects defined in a relationship. For example, containers have a varying cardinality of subobjects, ranging from zero for an empty container to unbounded for a resizable container.

**Child class:** is the immediately derived or descendant class in an inheritance relationship, and, as such inherits all the parent data and functionality but may access only that data and functionality that is public or protected.

**Class construct:** is used to define a type by specifying data fields (members) and member functions (methods). Essentially an ADT with encapsulation, the class construct distinguishes between public (external) and private (internal) accessibility to defined member data and functions.

**Class invariant:** is a documented summary of the properties, characteristics and functionality of a class – contains unique, unsorted elements with full support of copying. Under Programming by Contract, the class invariant specifies design details of interest to both the client and the class designer.

**Clean slate:** is a colloquial term that refers to software design that starts from scratch. That is, the software designer need not reuse, support, or integrate any existing code.

**Code bloat:** is the generation of excessively large amounts of code, often unnecessarily. Causes of code bloat include inappropriate optimizations (such as function inlining and loop unrolling), poor software design, and redundant instantiation of templates.

**Code complexity:** is a term used to describe how easy or difficult software is to read, understand, and maintain. Also known as software complexity, code complexity is not a performance measure.

**Code reuse:** is the use of existing software to build new software. Software libraries are well-known examples of code reuse: the utilities provided by libraries, such as I/O and pseudo-random number generators, are used over and over again by many different software systems. Code reuse may be more formally known as software reuse.

**Cohesion:** is a software engineering measure of functional or type integrity within a design. Cohesion describes how well a software entity (function, class, component) hangs or sticks together. The more cohesive an entity is, the less dependent it is on external entities, and, thus, the more maintainable.

**Compaction:** is the shifting of allocated memory to one portion of the heap, in order to reduce fragmentation and, thus, improve software performance. Like the reclamation phase of garbage collection, compaction is pure overhead.

**Compiler:** is the software that translates source code written by a software developer into assembly or machine code. Resolution by the compiler is typically called static (static typing, static binding) because it does not change at run-time.

**Composite Principle:** refers to practitioners' preference for composition over inheritance.

**Composition:** is the structure of a complex data type as defined by the composite of several data fields (members), where each data member is an essential element and provides some functionality. Composition models the has-a relation.

**Concrete class:** is a class definition that is fully implemented; all class methods are defined so that objects can be instantiated from the class.

**Constant:** is an identifier that does not need memory allocated because its value does not change. The compiler substitutes the constant value wherever this identifier occurs.

**Constructor:** is a special class method that is called by the compiler when an object is instantiated, removing the need for an initialize() routine. Constructors should set the object in a valid, initial state. Constructors return no value and have the same name as the class.

**Constructor Injection:** is a form of Dependency Injection used when a lifetime association is more likely since the resource is passed into the constructor. Error response is limited to using a default or throwing an exception because constructors cannot return error codes.

**Container:** is a data structure whose primary responsibility is to hold or contain data. Common containers include stacks, queues, and sets.

**Containment:** is a conceptual model of the holds-a relation. An object contains or holds one or more subobjects. The subobjects do not provide functionality and are not typically owned by the container.

**Contraction:** refers to an inheritance design that reduces (contracts) the inherited parent interface by suppressing (or NOPing) one or more inherited public functions.

**Copy Constructor:** is the constructor that initializes a newly allocated object by copying the state (value of all data members) of a passed object. If not defined by the class designer, the C++ compiler generates a copy constructor that performs a bitwise copy on all fields – a design landmine if a C++ class allocates heap memory internally. C# does not provide a copy constructor.

**Copy & Paste programming:** is a programming technique whereby functionality or structure is replicated by copying code from one portion of the software system to another. Highly susceptible to error, copy & paste programming should be avoided as it undermines software maintainability.

**Coupling:** is a software engineering measure of the degree of dependency between two software entities. Low coupling implies little dependency on external entities.

**CPU:** is the abbreviation for central processing unit, the computational core of a computer.

**Data corruption:** occurs when two or more handles (variables) unknowingly reference the same memory location. One handle can thus change the value of the memory location unbeknownst to the other handle.

**Deep copy:** refers to the allocation and initialization of separate memory when copying data. Safe but expensive, deep copies may be avoided by using aliasing. C++11 move semantics provide a safe and efficient alternative to deep copying by transferring ownership of memory from expiring temporaries.

**Default constructor:** refers to the constructor provided, by default, by the compiler when the class designer does not provide one. The default constructor takes no arguments. Hence, the term is often confused with no-argument constructor.

**Defensive programming:** is a style of programming in which no assumptions are made about correct usage of the software. Hence, the software tests many conditions, such as illegal input, or use exception handling in order to prevent errors.

**Deferred methods:** also known as abstract methods: methods declared in an abstract class interface but not defined. Definition (implementation) is deferred to derived classes.

**Delegate:** is an object that serves to provide functionality or services. Typically, a delegate is a data member composed within another object. If so encapsulated, delegates may be easily replaced or modified.

**Dependency Injection:** exposes internalized data members; clients 'inject' (pass) a dependency (resource) into a class via Constructor, Property, or Method injection. The class retains control because passed dependencies need not be accepted. DI supports testing and maintainability.

**Dependency Inversion Principle:** states that high-level abstractions should not depend on low-level abstractions. For example, a class definition should not depend on an encapsulated filename. High-level abstractions are stable; low-level are not.

**Derived class:** is a descendant or child class in an inheritance relationship. This older term is used with the term 'base'.

**Design Patterns:** are established solutions to reoccurring problems. A design pattern is general and reusable with expected costs and benefits. For example, several creational patterns address the need for virtual construction in a statically typed language.

**Destructor:** is a special class method in C++ that is called by the compiler when an object goes out of scope. It should be designed to release any resources (such as heap memory) held by the object but may also be used to update bookkeeping details. Destructors return no value and have the same name as the class, preceded by the special '~' symbol.

**Diamond Inheritance:** occurs in multiple inheritance when a child inherits from two parents that share a common grandparent, thereby causing redundant (two) copies of the grandparent component (one through each parent).

**Disassembler:** is a software tool that examines the executable (object code) of another program and, extracts a representation similar to the original assembly language code.

**Dual perspective:** describes two different views of the class construct: the client who uses the external, published interface; the class designer who defines the interface and implements the type, maintaining internal control of state.

**Dynamic binding:** refers to the run-time resolution of a function call. The compiler translates a dynamically bound function invocation to an indirect JUMP statement. Function resolution is postponed until run-time by using a virtual function table. Dynamic binding supports polymorphism and heterogeneous collections.

**Encapsulation:** is a key characteristic of OOD: the data members and associated functionality of a type are bundled together (encapsulated) in a class definition, thus promoting high cohesion.

**Exception:** is a hardware or software error that disrupts the execution of software. Exceptions can be named and 'caught' so that run-time errors are avoided.

**Exception Handling:** is a systematic response to exceptions. Errors so raised are processed and, if possible, normal execution resumes. Exception handlers are (small) pieces of code that execute when associated exceptions are raised.

**Explicit allocation:** is the direct allocation (acquisition) of heap-allocated memory via a run-time call, for example use of the new operator in C++/C#/Java.

**Explicit deallocation:** is the direct deallocation (release) of heap-allocated memory via a run-time call, for example use of the delete operator in C++.

**Extension:** is a pure form of inheritance where the child class preserves inherited functionality but also extends the functionality provided by the parent.

**Friend:** is a C++ construct that permits a class designer to selectively open up the class to external functions and classes. Any function or class declared a friend in classX has access to all the data and functionality of classX, even that declared private or protected. The friend construct is not symmetric, transitive or inherited.

**Function signature:** is defined by the function name and the number, type, and order of parameters.

**Functional decomposition:** also known as structured decomposition, breaking major tasks into lower level functions thus promoting readability and code reuse.

**Garbage Collection:** is the automatic reclamation of heap memory no longer in use (garbage) that removes responsibility for memory deallocation from the programmer but is an imperfect process. Executing software must pause for the garbage collector to run. Garbage collection is not controlled by the programmer and may, in fact, not ever be invoked for small or short-lived applications that use little heap memory.

**Generic:** functions and types definitions use a type placeholder rather than a specific type. For example, a generic swap routine swaps values of any type since the actions are the same regardless of type.

Generic containers, such as stacks and queues, are also common. When needed, the generic definition is instantiated with a type.

**Handle:** is a the means of accessing data stored in memory. A variable is a handle.

**Hard coding:** is a discouraged practice that uses literals (such as '3.14') rather than constants (such as `const float pi = 3.14`). Hard coding is not maintainable: if a value changes, all occurrences of that literals must be updated. In contrast, constant variables promote maintainability: if a value changes, (say, `const int limit = 314159`), the programmer need only update one statement – the constant variable.

**Has-a:** is also known as composition. A class has-a data member that provides essential functionality to the composing object. This relation is often preferred to inheritance because it affords more flexibility relative to cardinality, association, and ownership.

**Heap:** has multiple meanings: 1) data structure; 2) portion of memory in a program.

1) The heap data structure is a tree represented by an array, where A[1] represents the root node, A[2] represents the left child of the root, A[3] represents the right child of the root, .... For array element A[i], the left child is A[2*i], the right child is A[2*i + 1], and the parent is A[i/2]. Priority queues are often implemented via the heap data structure.

2) The (run-time) heap is a portion of program memory that is used for the dynamic allocation of memory. Heap memory provides much flexibility but incurs run-time overhead and can result in performance degradation if poorly managed.

**Heap Fragmentation:** occurs when free memory is scattered across the run-time heap and is available only in small blocks, causing the allocator to work harder to satisfy memory requests (thus degrading performance). Additionally, a memory request may fail if there is enough memory available but not in large enough blocks.

**Heterogeneous collection:** holds polymorphic objects – each object can be of any type in a class hierarchy.

**Holds-a:** is also known as containment, a class holds one or more data members but does not derive any utility from these subobjects thus implying a lack of type dependency.

**Identifier:** is a (user-defined) name that refers to a constant, variable, function, etc.

**Implementation:** of a class is the code that provides functionality and embodies the design decisions made with respect to internal structure and support for a defined type.

**Implementation invariant:** records design decisions, intent, and assumptions relevant to the implementation of the defined class, and is essential for maintenance.

**Implicit Allocation:** is the indirect allocation (acquisition) of memory, without explicit calls to the allocator. Either the compiler allocates memory, via stack frames, or memory is automatically allocated at run-time (as in dynamically typed languages like Python).

**Implicit Deallocation:** relies on garbage collection to reclaim heap-allocated memory. Programmers do not explicitly deallocate memory.

**Indirection:** refers to the ability to access memory indirectly, via a pointer variable, or to invoke a function indirectly, via a delegate or function pointer.

**Information hiding:** is an ideal in software design that implementation details are hidden so that the client does not become dependent on arbitrary implementation characteristics. Information hiding is difficult to realize because compilers need type information (size) in order to lay out objects correctly or appropriately request memory.

**Inheritance:** is a key OO relationship where a child class is defined in terms of its parent class, 'inheriting' data and functionality. Inheritance supports the is-a relationship a child object can serve as a parent object. Structurally, inheritance can be mimicked with composition but inheritance designs are essential for polymorphism.

**Inlining:** is a compiler optimization technique that replaces a function call with the body of the function to avoid the overhead of function call and return. Inlining can improve performance but can also lead to code bloat, ironically, decreased performance.

**Instantiation:** refers to the allocation and initialization of an object via a constructor call.

**Interface:** is the set of functions defined for a class (or component, or module) that may be delineated by accessibility: public (client); protected (descendants); private (internal).

**Interface invariant:** is a set of state conditions or constraints that must be met to support client expectations. The interface invariant is a contractual specification for the client.

**Is-a:** is an inheritance relationship where the derived class maintains the interface and functionality of the base class so the derived object is-a base object. Substitutability is possible when the is-a relationship is supported.

**Legacy code:** is existing software that continues to be used, despite newer technology or improved methodologies. Often such systems function for convenience of established users' needs. Typically, replacement cost is considered prohibitive.

**Lifetime:** is the length of time that a variable (piece of data) remains allocated. Note that allocation does not imply utility or access.

**Liskov Substitution Principle:** is the OOD principle that verifies the interoperability of (sub)types defined in a class hierarchy. The substitution of any derived class object in place of a base object supports heterogeneous collections.

**Literal:** is considered 'hard-coding' and thus not maintainable, for example '7', 'Hello'. Literals are not associated with memory, and thus cannot be modified,. For maintainability, constants should be used instead of literals.

**Method Injection:** is a form of Dependency Injection used when the utility of a resource is confined and is passed only into the method that uses it. Typically, the class does not retain any responsibility for the resource.

**Mutator:** is a class method that alters the value of one or more data members of an object. A mutator need not induce a state change. For example, popping an item off a stack may not change the state of the stack, unless a non-empty stack becomes empty.

**No-argument constructor:** refers to a constructor that takes no arguments. Ideally, this term should be distinguished from the default constructor which is the (no-argument) constructor provided by the compiler when the class designer fails to define any constructors.

**Node class:** is an intermediate class in a class hierarchy that inherits form (and possibly some functionality) and anticipates extension. A node class itself may be partially abstract.

**NOP:** stands for No OPeration and is an operation code (opcode) that indicates that no operation should be undertaken.

**Object-oriented design (OOD):** refers to software design that rests on the definition and use of objects, as well as the specification of appropriate relationships between objects.

**Object-oriented programming language (OOPL):** support OOD by providing the class construct and built-in constructs for inheritance and polymorphism. A software developer can thus easily define inheritance and dynamic binding, without using arcane constructs such as function pointers.

**Open Closed Principle:** specifies that a class should be open for extension but closed for modification, and is a key design principle of OOD.

**Operating system:** is the software, typically pre-loaded onto desktop computers, that handles basic tasks such as IO (input from keyboard, output to screen or file), scheduling processes, organizing files and directories, and executing applications.

**Operator overloading:** is the definition of class methods that can be invoked through a symbol, such as '+'. C++ fully supports operator overloading. Java does not support any operator overloading. C# selectively supports operator overloading.

**Orthogonal:** refers to entities that do not overlap so they can be treated separately. In a multiple inheritance relationship, if two parent classes are orthogonal, their interfaces have no common functions and thus do not confound design with ambiguity.

**Overloaded:** functions are functions which share a name but which are distinguished by different parameter lists. Constructors are commonly overloaded in class definitions.

**Overridden:** methods occur in class hierarchies when a derived class redefines the implementation of an inherited method. Overridden methods must have the exact same function signature as the method inherited from the parent (or base) class.

**Ownership:** refers to the handle (variable/object) responsible for a piece of data (another object). Ownership should be tracked in order to avoid memory leaks and data corruption due to unwarranted aliasing.

**Parametric polymorphism:** is another name for templated or generic code – code written with a type placeholder. When explicitly instantiated, a type is supplied and the compiler generates a copy of that class or function with the parameter type filled in.

**Parent class:** establishes the interface to be used by descendant via inheritance. There is no conceptual limit on the number of child classes that can derive from a parent class.

**Pointer:** is a variable that holds an address. Available in C/C++ but not C#, the pointer construct provides the programmer with the power of indirection and explicit aliasing.

**Polymorphism:** is the dynamic binding of method calls within the scope of a class hierarchy. All calls are dynamically bound in Java. In C++ and C#, a base class must specify a method as 'virtual' in order for it to be dynamically bound. A derived class may override an inherited virtual function, providing variant behavior. Virtual function invocations are resolved at run-time, so, a single (polymorphic) call may yield many (different) results.

**Portability:** is a measure of how easy is it to move (port) code from one platform to another.

**Postconditions:** are conditions that hold after a function finishes execution. By evaluating postconditions, a client can track state and thus ensure the legality of subsequent calls.

**Preconditions:** are conditions that should be met before a function executes. By satisfying stated preconditions, a client ensures correct execution of a function.

**Principle of Least Knowledge:** is a design guideline that promotes low coupling by stating that one object should know as little as possible of another.

**Priority Queue:** is a queue that provides the same interface as a standard queue but internally orders items by priority, not in FIFO order. If data is low priority, then when queued in a priority queue, it may be stored indefinitely, that is, starve.

**Private:** confirms the encapsulated nature of class data members and methods. Any method or data member declared to have private accessibility cannot be accessed by either the client or descendant classes.

**Profiler:** is a software tool that runs code and analyzes execution, tracking memory usage and function coverage. Profilers can evaluate heap fragmentation, identify memory leaks and assess the frequency of function calls.

**Program counter:** holds the address of the currently executing instruction (in a special register) which is pushed onto the run-time stack when a function invoked so that, when the function terminates, control returns to the caller.

**Property Injection:** is a form of Dependency Injection as known as Setter Injection because the resource is passed into a set method that serves only to accept the dependency. Replacement and postponed instantiation can thereby be supported.

**Protected:** accessibility allows access for descendants. Any method or data member declared to have protected cannot be externally accessed by the client: private to the outside, public to descendants.

**Public:** denotes opens access to all. Any client or class may access public data and methods. Public data members violate encapsulation and are thus discouraged.

**Pure virtual:** refers to a C++ method declared but not defined. Also known as deferred or abstract methods, pure virtual functions make a class abstract. Inheritance is required: descendant classes must provide implementation details.

**Queue:** is a standard container that stores data in a FIFO (first-in, first-out) order. Enqueueing adds to the back of the queue; dequeueing removes from the front of the queue.

**Raw pointer:** refers to the C/C++ pointer construct which serves as an address holder, and, unlike smart pointers, does not implicitly ensure resource deallocation.

**Readability:** refers to the ease of reading and understanding code, and thus reflects maintainability. Code construction guidelines suggest techniques such as functional decomposition, encapsulation, and self-documenting code to promote readability.

**Redundancy:** occurs when a child inherits from two parents that share a common grandparent. The child class object thus receives two copies of the grandparent components. C# and Java do not support multiple inheritance and thus do not encounter this problem. C++ designs may avoid such redundancy through virtual inheritance.

**Reference:** is a variable that holds the address of data in memory. Multiple references can address the same memory, thereby establishing aliases and supporting sharing. Poorly tracked aliases (references) may lead to data corruption.

**Reference Counting:** associates a reference count with each allocated memory block. Every reference to a block increases its reference count. Each time a reference is reassigned, or goes out of scope, the count is decremented. A count of zero indicates that there is no access and so the memory block may be reclaimed.

**Requirements:** are specifications that define the essential functionality of a software system and may include non-functional characteristics (such as performance).

**Responsibility Driven Design:** is a principle that stresses clear identification of class functionality and dependencies.

**Root set:** is the set of variables in scope when program execution is paused so that the garbage collector may run. A trace emanating from the root set identifies all active variables so that the garbage collector will not reclaim memory still in use.

**Scalability:** is a measure of how well software performs under increased load conditions.

**Self-documenting code:** is the deliberate selection of identifier names that describe use and intent. Variable names such as min and max clearly imply intent as do functions calls like Fibonacci(n).

**Shallow copy:** establishes an alias (a secondary reference) to a piece of data in order to avoid the overhead of allocation and initialization of a true copy. Efficient but vulnerable, shallow copies may lead to data corruption.

**Side Effect:** is an unintended or secondary effect of a direct action, such as a function call that alters persistent data without the caller's knowledge.

**Single Responsibility Principle:** is the design principle that prioritizes the isolation of primary functionality in as class design without inclusion of secondary functionality.

**Smart pointer:** is a wrapped pointer that serves to guard against memory leaks in C++. When a smart pointer goes out of scope, its destructor is invoked so that any heap memory referenced by the smart pointer is appropriately deallocated. C++11 provides three types of smart pointers: unique, shared, and weak.

**Software complexity:** assesses software's structure, readability, and maintainability. How intricate, layered, complex is the software? Common measures of software complexity include control flow, coupling, branching, data, data access, and cyclomatic complexity (the number of independent paths through the software).

**Software Engineering (SE):** applies engineering principles to developing and managing software systems; includes requirements analysis, design, implementation, testing, maintenance, and reengineering.

**Software Evolution:** is a modern term referring to adaptive software maintenance, the upgrading of existing software in order to provide more functionality, improved performance, etc.

**Software Maintenance:** is the modification or upgrading of software. When viewed as corrective, maintenance implies fixing defects or improving performance. Yet, support of an expanding software system includes functional enhancements, refinements of UIs, platform extensions, etc.

**Source code:** is a set of executable instructions, usually written in a high-level language.

**Specialization:** is a form of inheritance where the derived class modifies or extends the parent functionality in a manner that specializes the behavior according to subtype.

**Specification:** is a form of inheritance where the derived class provides implementation that is missing in the abstract parent class.

**Stack:** has multiple meanings: 1) data structure; 2) portion of memory in a program.

> 1) The stack data structure is a common container that stores data in a LIFO (last-in, first-out) order. Its classic interface supports pushing (storing) and popping (retrieving) items.

> 2) The run-time stack is a portion of program memory that holds data currently in scope. Each function call causes an activation record (stack frame) to be pushed onto the run-time stack. Upon function exit, the activation record (or stack frame) is popped off the run-time stack. Hence, recursive calls effectively hide the provision of local data with each call and unbounded recursion leads to stack overflow.

**Stack frame:** is also known as an activation record: a layout of the data needed to process a function call and thus includes the program counter as well as space for local variables.

**Starvation:** is a possible side-effect of using a priority queue: a low-priority item may be stuck in the back of the queue as higher-priority items are enqueued ahead of it. Use of an internal aging mechanism may be used to avoid starvation.

**Static Binding:** refers to the resolution of function calls by the compiler whereby a function invocation is translated into a direct JUMP (to the function address). Static binding is efficient but not flexible. There is no overhead incurred at run-time to process the call but the address in the direct JUMP statement cannot change.

**Static method:** refers to functions declared and defined within class scope but not accessible via an instantiated object. Invocation is through the (scope of) the class name.

**Static variable:** refers to data members declared and defined within class scope not accessible via an instantiated object. When a class definition is loaded, one instance of the static variable is allocated. Every object instantiated from the class definition thus shares this one copy.

**STL (Standard Template Library):** for C++, provides generic versions of standard data structures, such as stacks and vectors, as well as standard algorithms.

**Structured programming:** promotes functional decomposition and use of control structures, and thus is often heralded as the emergence of software design, or as a constructive response to overuse of GOTO.

**Substitutability:** reflects the is-a relationship: a derived object can substitute for a base class object.

**Subtype Polymorphism:** rests on inheritance and dynamic binding. A derived class can override (redefine) an inherited virtual function so that, at run-time, either the base or a derived class method is invoked, dependent on the (sub)type of object through which the method is invoked.

**Syntactic Sugar:** is a derogatory term that implies that a language construct does not provide significant additional design support but merely sweetens the code.

**Templates:** are the generic type and generic function support in C++. A type placeholder is used in a template class or function definition. The compiler fills in the type when the programmer instantiated a template with a specified type.

**this pointer:** is the address of the object through which a class method is invoked; the compiler patches in this address as an implicit parameter of the call. Class methods can then reference data members associated with one specific object so data access is facilitated while data integrity is preserved. Static class methods are called through the class name and thus do not have a this pointer as an implicit parameter.

**Type extension:** is considered a pure form of inheritance: the definition of a new descendant extends the type definitions provided in a class hierarchy.

**Variable:** is a data identifier associated with memory; the value held in a variable may change (vary).

**Virtual function:** is a function tagged as virtual in its class definition, or overridden in a descendant class, so that it can or will be dynamically bound.

**Virtual function table (vtab):** is a table of function pointers associated with a class. Each entry contains the address of the corresponding (virtual) class method. When a function is defined, its address is placed in the table. The vtab entry of an undefined (abstract) method is zero.

**Virtual Inheritance:** is a tagged definition of inheritance in C++ that attempts to resolve redundancy possible with multiple inheritance.

**Wrapper:** is a class that wraps up, or encapsulate, an existing class. Wrappers typically facilitate code reuse by adjusting interfaces while retaining existing functionality.

# Bibliography

Blaha, M., Rumbaugh, J., *Object-Oriented Modeling and Design with UML*, second edition, Prentice-Hall, 2005

Budd, Timothy, *An Introduction to OOP*, 3rd edition, Addison-Wesley, 2002

Bulka, D., Mayhem, D., *Efficient C++: Performance Programming Techniques*, Addison-Wesley, 1999

Dingle, A., *Software Essentials: Design and Implementation*, CRC Press, 2014

Dingle, A., Hildebrandt, H., *C++: Memory First*, Franklin, Beedle & Associates, 2006

Ellis, M., Stroustrup, B., *The Annotated C++ Reference Manual*, Addison-Wesley, 1990

Gamma et al., *Design Patterns*, Addison-Wesley, 1995

Hildebrandt, Thomas, private communication, 1999

Loshin, David, *Efficient Memory Programming*, McGraw-Hill, 1999

Meyers, Scott, *Effective Modern C++*, O'Reilly, 2014

Seacord, R., *Secure Coding in C and C++*, second edition, Addison-Wesley, 2013

Stroustrup, Bjarne, *The C++ Programming Language,* special edition, 2000, Addison-Wesley

## QUOTES

defprogramming.com
google.com
quoteinvestigator.com
reddit.com
wikiquote.org

# Index

NOTE: Locators in *italics* represent figures and **bold** indicate tables in the text.

Printed in the United States
by Baker & Taylor Publisher Services